Communicating Science and Managing the Coronavirus Pandemic

V J MARCHESANI, PHD

PAGE PUBLISHING
Conneaut Lake, PA

First originally published by Page Publishing 2022

ISBN 978-1-6624-8021-8 (pbk)
ISBN 978-1-6624-8022-5 (digital)

Printed in the United States of America

To my wife, Rosalie, for her continuing love, patience, and understanding—without which, this book would not have been written

To my children, Darlene, Judy, and Keith, for their love, support, and kindness

To my grandchildren—Emily, Andrew, Mariella, Matthew, Abigail, and Luke, for their love, care, and encouragement

CONTENTS

List of Figures..vii
Author's Note..ix
Chapter 1: Introduction..1
Chapter 2: The Role of Science and the Acceptance of
            Science during a Pandemic....................................5
Chapter 3: The Use of Mathematics to Support Science 16
Chapter 4: Are You and the World Prepared for a
            Global Pandemic?..31
Chapter 5: How Do You Review the Risk and Prevent a
            Pandemic? 45
Chapter 6: Communications during a Pandemic 54
Chapter 7: The Difficulty and Importance of Testing
            Science to Ensure the Communication of
            Accurate Information 80
Chapter 8: Behaviors Prior to, During, and After a Pandemic....98
Chapter 9: The Event (Cluster of Illnesses) and
            the Criteria to Determine a Pandemic,
            Epidemic, or Crisis..109
Chapter 10: The Nine Stages of Managing a Pandemic..............127
Chapter 11: Developing or Improving Plans to Address
            the Pandemic and Spin-Off Crises........................142
Chapter 12: The Activities of the Pandemic Management
            Team during a Pandemic....................................157
Chapter 13: The Fundamentals of Managing a Pandemic..........171
Chapter 14: The Use of a Crisis Management System................181
Chapter 15: Pandemic Stability, Remediation, Recovery,
            and Closure..189

Chapter 16: The Local Community, Country, and World
Community and a Pandemic Situation....................196
Chapter 17: The Media and the Crisis Situation Rooms.............201
Chapter 18: The Pandemic, Crisis Scenario Creation, the
Exercise, and the Critique.................................214
Appendix 1: Questions at Time of a Pandemic.......................227
Appendix 2: List of Resources 229
Appendix 3: Sample Standby Statement 231
Appendix 4: Pandemic and Spin-off Crises
Management Plans......................................233
Appendix 5: List of Organization Documents 245

List of Figures

Figure 1:   Heads and tails coin distribution forming a
            bell-shaped curve 20
Figure 2:   Bell-Shaped Curve from Dice 25
Figure 3:   Public Confidence 72
Figure 4:   Flow Chart 148
Figure 5:   Declaring a Pandemic Severity and Probability 173
Figure 6:   Declaring a Pandemic Multiple Spin-Off Crises........175

Author's Note

The following text does not guarantee or warranty that what follows will prevent an individual or organization from experiencing a pandemic or spin-off crisis situation. The text presents thoughts and ideas that may better prepare an individual or organization for such situations.

CHAPTER 1

Introduction

Communicating science is often difficult and sometimes almost impossible. Learning about science is a building process. Early on there is a general science curriculum followed by biology, chemistry, physics, and other more detailed science courses. Science is often accompanied by mathematics in the form of arithmetic, algebra, geometry, calculus, and other more detailed mathematics topics. For people without science and mathematics backgrounds, understanding something like virology, the study of viruses such as COVID-19, is difficult and has its challenges.

Years ago, I, as a scientist, took a communications course. The professor gave an assignment to write a paper describing the scientific term *pH*. The assignment seemed straightforward; *pH* was a term of science that could be defined. However, the professor added, the paper that the class was to write is to be read and understood by a six-year-old child.

This changed everything! *pH* is defined as minus the log of the hydrogen ion concentration. There was no way a six-year-old child nor, for that matter, anyone without a science background would understand the definition. I had my work cut out for me. I realized this assignment was not about me showing my knowledge of science or how smart I am or not. I was to find a way to reach a six-year-old child that would keep their attention, and more important, they would understand what pH is and why they should want to know about it. The assignment taught me a great deal. Also, it is important to understand that a scientist is not all-knowing and aware. If a scientist cannot communicate their knowledge in a simple and effective way, they cannot share what they know, and that is a big problem.

You may think you are not capable to understand science; therefore, you cannot trust science. The problem is not you; the problem rests with science. Science needs to do a better job communicating. Therefore, one of the areas this book addresses is communicating science during a pandemic, or how to take complicated information, data based on science, and provide that information in an easily understood way that it can be accepted and followed.

In 2019, 2020, 2021, and possibly 2022 and beyond, the world faced and faces a pandemic health crisis. A pandemic occurs when a virus is prevalent and impacting throughout the world. This pandemic began in late 2019 with the identification of a new virus, Coronavirus 19, or COVID-19. The name Coronavirus 19 means it is a coronavirus and was first identified in 2019. There was no COVID-18, 17, 16 and the like. The virus fulfilled its potential by spreading rapidly throughout the world, resulting in a global pandemic. While most common colds are from coronaviruses, this coronavirus was new. Make no mistake, a pandemic is a health crisis of global proportions.

In this book, I shall step through topics such as: Is and was the world prepared for a global pandemic such as COVID-19? How do you review the risk and prevent a pandemic? What are the behaviors prior to, during, and after a pandemic? What are the nine stages of managing a pandemic and the criteria to determine if a pandemic exists? Developing and improving a plan to address the pandemic, the activities of your pandemic management team; the fundamentals of managing a pandemic; the use of a crisis management system plan do check and act in addressing a pandemic; pandemic stability, remediation, recovery, and closure in the local and world communities. The role of science and the acceptance of science in a pandemic situation, the media in a pandemic situation, the development of a pandemic scenario to test pandemic response preparation, a pandemic exercise and critique, a pandemic management flow chart, pandemic matrix, and multiple levels of involvement pandemic matrix will also be included.

In addition, I shall provide questions to be asked at the time of pandemic, a pandemic resource list, an example of a pandemic standby statement, and a template for a pandemic management plan, an example of documents to be used at the time of a pandemic, message development, and a scripted pandemic exercise. Bill Gates, in a TED Talk in 2015 titled, "The Next Outbreak, We Are Not Ready," pointed out that the World Health Organization (WHO) monitors world health, looking for epidemics and pandemics. Their role is to find them, and the world collectively and individually needs to respond. Mr. Gates pointed out, "We need to be prepared." Further, he made important recommendations as he pointed out that a pandemic has the potential to have a greater impact than war. He suggested that we need to create strong health systems in poor countries simply because these areas are very vulnerable to an onset of illness that may lead to a pandemic. He identified a need for a core medical reserve to move quickly to areas showing early signs of a pandemic. He recommended that the military be paired with medical response teams because the military knows how to move people and equipment quickly across great distances. He suggested simulations to test response systems. He explained that an attempt at a simulation in response to a pandemic was a failure. This should have set off alarms. Lastly, Mr. Gates suggested lots of research and development.

One word that I am not hearing in this pandemic and its spin-off crises is the word "cure." This should be concerning. The absence of the word "cure" implies that illness from COVID-19 cannot be cured. We need to take a closer look. If I have an illness caused by a bacterium and take an antibiotic, I expect to be cured. The antibiotic impacts the biochemistry of the organism that is causing me to feel sick, and as a result, the organism dies, and I feel better, and I am cured.

Viruses do not work this way. They are not considered alive, and they cannot reproduce. Unlike bacteria, there is no biochemical pathway to be interrupted causing an organism to die. Viruses are not organisms. Vaccines tend to block entrance of the virus to the body cells. Other approaches are to take steps to lower the probability of the virus from entering the body. Social distancing, the use of quarantine of individuals who were in contact with someone who has COVID-19, the use of masks, contact tracing, the use of gloves among other choices to keep the virus away. There are treatments once people have COVID-19. These include monoclonal antibodies, a new pill treatment, and other treatments. These treatments are designed to have the body impacted by COVID-19 replace the cells containing the COVID-19 virus with healthy cells, resulting in the patient feeling better. The use of monoclonal antibodies needs to occur in the first twenty-four to forty-eight hours after testing positive for COVID-19. Beyond that time, the treatment does not appear to be as affective. Like the 1918 influenza, virus where we today are asked to take a flu shot every year, we should expect the COVID-19 virus to be around for many years. As with the influenza virus, we shall learn to live with it and take the appropriate precautions each year.

CHAPTER 2

The Role of Science and the Acceptance of Science during a Pandemic

The scientific method

First and foremost, I want to begin with a discussion of the scientific method. This is taught to students at the outset of any science studies. The method is extremely important to the understanding of the thinking of scientists. It also addresses how scientists may prepare to communicate findings to the community.

The method refers to a specific process designed to identify and evaluate a finding to provide knowledge, understanding, and hopefully how to communicate a path forward. The process steps begin with an observation. For example: observing a sunset. This is followed by asking a question such as, "Why does the sun set, and what causes it to happen?" The scientist develops reasoning as to why he or she believes the sun sets, a hypothesis; but the scientist must be able to test their reasoning to determine if it is correct. For example: by gathering information such as, the sun is a fixed object in the sky and does not move. As a fix object, it does not set. Further, it is learned that the earth rotates on its axis every twenty-four hours. It is the earth rotating on its axis that causes the sun to appear to set. With this knowledge, the scientist can test their hypothesis concerning sunsets by reviewing literature and learning information.

In addition, the light from the sun travels at a very rapid speed, 186 thousand miles per second, and the sun is 93 million miles from earth. Therefore, it takes about eight minutes for a particle of light—yes, I said a particle of light—to leave the sun and arrive on earth. This makes things interesting, Therefore, since it takes eight minutes for the particle of sunlight traveling at 186 thousand miles per second to arrive at the earth; when you observe the sun setting, what you are seeing occurred eight minutes prior to you seeing it. Another interesting point concerning the sun is that it takes the earth 365 days and 6 hours to travel around the sun. We identify the time the earth travels around the sun as a year, 365 days. However, it takes 365 days and 6 hours for the earth to go around the sun. Every four years, the six additional hours for the earth to travel around the sun are added together for twenty-four hours, a day. This is added to the calendar; and we call this 366-day year, leap year. I have chosen to include the information above to explain that science can be effectively communicated in an understandable manner.

Sunset

The scientific method can also be, and is, used to test conclusions. The method helps provide the documentation needed to draw conclusions based on science, not opinion. It is a path toward learning truth. As you may imagine, the scientific method takes time. So, while people wait for answers from science, they often get frustrated as to why the scientist cannot tell them all is well; that is all they want to hear. For example: they want to hear that the COVID-19 vaccines are the answer to normalcy.

Often variables, things that interfere, determine the extent of accuracy. Therefore, variants in the COVID-19 world must be studied and evaluated to determine their possible impact and whether existing vaccines will prevent someone coming down with COVID-19 from the variant. Yes, we are all looking for that exhale that says all is well; but that exhale will feel much better if based on science and not on unsupported rhetoric.

Truth

All this leads to the question: What is "the truth?" or is science providing the truth, and can we trust it? Can we believe in science, and why? This question is as difficult as, "What is the meaning of life?" We all are in search of the "truth," but too often it is difficult to find. The search for truth begins with a belief. Because you believe something is true does not mean it is true. Many people believe truth is reached when a group of people have the same belief. In other words, there is a consensus that something is true; therefore, it is true. For example: There is a group of people who believe that the world is flat. If truth is based on group consensus, then the group that believes the world is flat, when gathered, would be correct. Since there is an abundance of documentation, e.g., pictures, and more proving the world is not flat, then group consensus obviously does not determine truth. Hardliners of the-world-is-flat thinking will not accept documentation and continue with the incorrect position that the world is flat. One wonders that their persistence in their belief in the wake of overwhelming evidence is nothing more than a desire for notoriety or recognition. To consider something as true, you need something to support that truth well beyond consensus or even a majority. You need exactness. The more exactness of documentation, the stronger the truth. Often you will hear people say, "That is absolutely true." In other words, what is being said has no qualification, restriction, or limitation. In essence, the person commenting is saying, "What I have said is the truth and cannot be challenged!"

Mathematics in the form of statistical analysis is often used to test conclusions. Statistical analysis can shut down a claim or conclusion or support it to the point that the claim or conclusion cannot be challenged. Only the statistics can be challenged, and assuming the arithmetic is correct, the statistics cannot be challenged. This will be further discussed later.

Why 5X

Another method to find the truth or get close to the truth is to ask *why* five times. If the person responding to the question of why five times does not revert to lying, you will get close to the truth. For example: your friend Fred sells his car, and you ask the following, five "why" questions:

Question:
   Why did you sell your car?
Answer:
   I wanted a new car.
Question:
   Why did you want a new car?
Answer:
   My car was old and did not look nice.
Question:
   Why do you think it did not look nice?
Answer:
   Most of my friends have cars that look better.
Question:
   Why is that important to you?
Answer:
   I do not want my friends to think less of me.
Question:
   Why is what your friends think important to you?
Answer:

Well, I like Mary, and I want to date her, and I thought she might not go out with me because my car is in such bad shape.

The truth, or closer to the truth, Fred sold his car to impress Mary who he wants to date.

Again, asking why five times often gets you closer to the truth.

Unfortunately, there is a major need and effort underway to get people who are not vaccinated, vaccinated. Data forthcoming from hospitals around the United States identify the fact that presently 80 to 90 percent of the people in the hospital with COVID-19 are unvaccinated. It is interesting that there are several comments from people who won't get vaccinated. I have listed only some of the comments and my responses to their comments.

- Comment: "I do not trust science."

  Response: What is missing is, why do you not trust science? Science is difficult, and you cannot learn it in an afternoon. Many people do not want to spend the time learning science. It is much easier to say, "I do not trust it," rather than spend the time learning about it. In addition, no one wants to get embarrassed by saying the words, "I do not understand."

- "A friend of mine told me the vaccine was not safe."

  Response: You need to ask the question, "Why do you say the vaccine is not safe?" If your friend responds that he heard this on television, that is not considered documentation supporting the comment that the vaccine is unsafe. You need documentation that you can hold in your hand and read and understand. This documentation should include the data that clearly make the case. To date, this type of confirmatory documentation discrediting the vaccine has not been presented.

- "Heard the vaccine will cause me to get magnetized such that a key will stick to my forehead."

  Response: This comment is wrong and irresponsible. It would imply that the vaccine contains metals found in magnets. Nothing has been presented that supports this comment.

- "I never liked science in school, so why should I begin to believe it now?"

  Response: Not liking science is no reason to reject the result generated by science. Time must be taken to understand the science, the generation of results, and understanding of the comments of the expert interpreting the study results.

- "A friend of mine had COVID-19, recovered from it, and he did not get vaccinated."

  Response: Your friend is fortunate to have recovered from COVID-19. Having recovered from COVID-19 means that your friend has antibodies having recovered from the COVID-19 virus. The antibodies will help to prevent your friend from getting the virus again should they be exposed. The problem is that it is unclear how long the antibodies are effective in preventing the disease from returning. Your friend is far better off waiting a period and getting vaccinated with a vaccine that has an expected efficacy of about eight months before getting a booster shot. Your friend should consult their physician for the time they should wait to get the vaccine after recovering from COVID-19.

- "I believe in my freedom, and I should choose if I want the vaccine or I want to wear a mask."

  Response: Yes, you are intitled to your freedom, but you are not intitled to have and spread the COVID-19 disease. There is a difference between your freedom affecting you and your freedom affecting others. You're certainly entitled to where your freedom affects you; but you're not entitled where your freedom affects others. True freedom is where you are able to function as you do without the presence of the pandemic. The evidence is clear that vaccinations and mask-wearing are precursors to true freedom. We need herd immunity to take place with the COVID-19 virus. Herd immunity will take place when some 80 to 90 percent of the population has been vaccinated. People who are not getting vaccinated to bring us closer to herd immunity are prolonging the pandemic.

- "Children are not affected that much by the virus, so there is no problem for schools."

  Response: This was the thought process early in the pandemic. The view was, the children are minimally affected or not affected at all. As the pandemic has gone on, and with the introduction of the Delta variant, children are impacted. Those children that are under the age of twelve can now be vaccinated. This along with social distancing and mask-wearing to will help to prevent the impact of COVID-19 and its variants in children.

- "I heard COVID-19 treatments work, and I plan on getting them if I feel sick, for example, the monoclonal antibody treatment."

  Response: Monoclonal antibodies is the treatment provided ex-president Donald Trump when he came down with COVID-19. The treatment requires that it be administered early within twenty-four to forty-eight hours after being diagnosed with COVID-19. Those that are objecting to the vaccine and what it contains and that it was developed within a nine-month time frame should look at what the monoclonal antibodies contains and the time for its development.

- "A parasitic treatment seems to have some merit, and I want to learn more about it."

  Response: COVID-19 and its variants are not parasites. Parasites have been researched and are well understood. Since COVID-19 is not a parasite, one wonders why someone would be willing to take this so-called treatment with any degree of confidence.

- "The CDC speaking about science is giving us mixed messages. I do not know what to believe."

Response: Messages change because things change. As more information is learned about Covid 19 and its variants, actions to prevent disease change. As the case numbers came down for those who had received both shots of the Pfizer or Moderna vaccines, the CDC announced that use of masks was unnecessary. With the presence of the Delta variant, the CDC has reconsidered and suggested the use of masks is appropriate under certain circumstances. Since 80 to 90 or better percent of COVID-19 and its variants enter the body through the nose and mouth, one might wonder why not wear a mask to prevent the virus from entering the body or in the case where someone has COVID-19 to prevent it spreading.

- "I listen to what some are calling misinformation about the vaccine; however, science does not tell me what I am hearing and reading are wrong."

  Response: This is a very important point. Where science does not come down hard and fast shooting down misinformation, it is implied that the misinformation is correct. Nothing could be further from the truth. Science must take the time to address each form of misinformation. The misinformation is harming and, in many cases, killing people. Science must speak out with authority, as it can, to stop this incorrect information when it first appears in any form.

- "My government leadership is not mandating me to get vaccinated or wear masks, and I like that."

  Response: In many cases, there are government mandates to protect the health and well-being of the population. Before children can enter school in the United States, they must be vaccinated for polio, mumps, and measles. I hear no one standing up and saying this mandate is against the freedom of their children. What is missed here is, young children, if they understood the importance of these three vaccines, would say, "I choose the vaccine over dying or being crippled for the rest of my life." In addition, there are government mandates in many states, for example: if you drive a vehicle that vehicle must be inspected so that there will not be brake failures or other mechanical malfunctions of the vehicle that would hurt the driver his or her passengers, pedestrians, or people in other vehicles. There are many more examples of government mandates including taxes among others that people do not stand up and claim that their freedom is being impacted. The government is expected to soon issue vaccine mandates for all industry employing more than one hundred people.

- "Our government is the problem, giving me wrong information, not telling me the truth."

    Response: The answer here is simple. If you believe that the government is not telling you the truth, then request documentation of what government is saying. If documentation cannot be provided, then you have every reason to question whether what you are hearing is the truth. In the same way you must require documentation from those providing misinformation and conspiracy theorist who speak information and do not document it. Therefore, their comments must be questioned. Items that are presented with the words "I heard" or "I recently read" or "My friend told me," do not carry weight. Documentation to support their comments must be provided or their comments need to be disregarded.

- "Too many voices are saying something different. I do not know who to trust."

    Response: I agree there are too many voices saying different things. Which one should be trusted? The answer is easy. The one that provides the documentation that is supportive of the comments. The documentation must be supported by science and math. If that is not the case, disregard what's being presented or said.

- "I heard someone got vaccinated, and their arm fell off."

    Response: This is an example of a clear case of misinformation. I heard this on television. If you did not see any concrete evidence/documentation supporting the statement, please discard the misinformation.

- "Now I hear people who got vaccinated are getting the virus. Therefore, getting the COVID-19 virus can take place whether I am vaccinated or not."

    Response: Yes, the vaccination provides information to the body which causes the development of COVID-19 antibodies. It is unclear how long these antibodies last to protect the vaccinated from COVID-19 or its variants. It is estimated that the antibodies created by the vaccine last some eight months after the second dose of the vaccine. For this reason, booster shots have been identified. The booster shots will cause the development of additional antibodies that the body creates that will protect you. You may ask, "How long will the booster dose protect me?" The answer is not currently known because there are no data available. From experience to date, one would expect this third dose to last at least eight months.

    For those who have had COVID-19 and survived, you have generated antibodies without the vaccine. However, it is recommended that those who have survived COVID-19 get vaccinated about three months after they have recovered from COVID-19. This is because it is not known how long you will be protected with the antibodies generated from having the COVID-19 virus.

- "I hear the flu season is coming shortly. I do not get the flu shot. I guess I shall be asked to get the two COVID-19 vaccine shots and a vaccine booster, and a flu shot. No, I'm not doing any of that."

    Response: Refusing something that is documented to protect your health is foolish. The decision not to get vaccines is not a badge of courage. It is an unnecessary risk that is being taken.

For those of you who get the flu shot every year, the flu shot is designed to prevent you from getting the influenza virus, which was the virus of the 1918 pandemic. The influenza virus of 1918 still exists. It is it for this reason that it is highly likely, like the 1918 influenza virus, the COVID-19 virus and its variants will be around for a long time. Simply stated, viruses cannot reproduce and therefore are not alive and can exist for extended periods of time. For example: the influenza virus has been around for more than one hundred years.

Many people, no matter what they're told, will not take the vaccine. One needs to ask the question, why? There is no one-size-fits-all answer, but we can make some assumptions.

- o They do not trust science nor the messenger and will not take the vaccine.
- o They are in with a group of friends, safety in numbers, where they are afraid they will be rejected by the group if they receive the vaccine. Some people have been identified as wearing disguises to get vaccinated so not to be embarrassed when and if their friends learn of their decision to get vaccinated.
- o "I'm in good health, and there's no reason for me to take something into my body that I don't trust." Being in good health does not mean anything to the virus. You still can be severely impacted and die.

The critical point here is that science must do a better job in communicating reality and factual information that may easily be understood. People need to know the facts! They do not need jargon! Facts about this virus and its variants must and can be communicated in a straightforward, easily understood format.

CHAPTER 3

———

The Use of Mathematics
to Support Science

Science and Mathematics

Before running away or skipping what follows, I think we need to understand certain principles. Science and mathematics are linked. They are married to one another; each makes the other better.

The sciences depend on mathematics to help explain what our senses tell us. It is from the mathematics that we learn what needs to be done to improve or correct a situation, and it improves our problem-solving skills. Therefore, mathematics is critical to our life and to science. For example: the science of meteorology includes numbers such as barometric pressure, temperature, and relative humidity. If we wish to know the average daily temperature for the summer months in New York City, we would add the hourly temperatures for each day during the summer and divide that resultant number by 24 to get the average daily temperature for the summer season in New York City. We could also add the average daily temperatures for each day during the summer season and divide that number by the number of days in the summer season to obtain the average daily temperature during the summer in New York City. Either method should provide the same result. The reason for this simple analysis is to demonstrate the use of scientific data married to mathematics to generate accurate information. Information that is far better than someone identifying the average daily temperature as a number someone told them, but they do not know that person's name nor where they learned the number, but they are a knowledgeable person. With mathematics to help document the average daily summer temperature in New York City, there is a strong basis for acceptance; while the number identified by a person with no name or documentation has no basis for acceptance and the number should not be used. It is obvious that knowing the average daily temperature in New York City is not a critical item. However, if the example addressed the number cases of COVID-19 for nonvaccinated people, you would want accurate data, and not someone saying, "There is no problem here," with no documentation for the comment. Support documentation is a must if accurate information is needed. Conspiracy theorists provide information based on undocumented information. Such information is dangerous. There is a need for demanding documentation for every COVID-19 comment. Without it, the documentation, the comments need to be rejected and discarded.

One interesting item is that most people speak about the high humidity during the summer. Humidity is reported as relative humidity, not humidity even though you may hear the weatherperson referring to relative humidity as humidity. So why is it interesting? Humidity refers to the amount of water in the air. When you hear the relative humidity is 100 percent, it is raining. The interesting part is that when the atmosphere has a water concentration of 3 percent, it rains. We can experience 100 percent humidity when we are below the surface of water. So relative humidity refers to how close we are to 3 percent water in the atmosphere. One hundred percent relative humidity is equal to 3 percent moisture in the atmosphere, and it rains. Here again science and mathematics play an important role in providing accurate and correct information.

Statistics

Scientists use statistics to evaluate collected information, better known as data. Remember data is plural, datum is singular, and data do not speak. So, the next time you hear someone say, "The data says," please understand that since data do not say anything, what you are hearing is someone's, hopefully an expert's, interpretation of the data. Please note that I did not say opinion. I said interpretation. Statistics means many different things to many people. Some believe you can influence statistics to provide information that permits you to draw any conclusion you choose to reach. To begin statistical analysis is used to evaluate data. It is the interpretation of the data/statistics by an expert who is both educated and has experience that permits data interpretation and the drawing of conclusions. I shall try to stay away from statistical jargon in trying to make statistical analysis more meaningful and understandable. In other words, I shall provide an example of how statistical analysis works.

If I were to take ten United States minted pennies and tossed them into the air when they all came down to a surface, table, floor, whatever, some would be showing the head of President Lincoln, typically called "heads" and the other, the back side of the penny coin, typically identified as tails. In looking at the coins that have landed, I can ask the question, how many heads are visible, and how many tails are visible? The count can be any of the following:

- ten heads
- nine heads and one tail
- eight heads and two tails
- seven heads and three tails
- six heads and four tails
- five heads and five tails
- four heads and six tails
- three heads and seven tails
- two heads and eight tails
- one head and nine tails
- ten tails

Sample distribution after tossing ten pennies

This is simple and straightforward. The question for the statistician is, "How often, when I toss the coins in the air, should I expect each of the above distribution of coins?" Since each coin is two sided, there is a 50 percent chance when the coins settle after being tossed into the air of displaying a head or a tail. It is highly probable that when the coins settle, at least one will be a head or tail. It is also highly probable that of the ten coins two, three, four, or five will be a head or a tail. The idea of all being heads or all being tails has the lowest probability. So why is this important, and what does it mean? If we took the time to toss the coins into the air one thousand or more times and counted the number of heads and tails each time, we would be able to create a table that displays the results of tossing the coins in the air one thousand or more times. We could then convert the results into a graphic form called a curve. The curve would resemble a bell with the highest combination of coins being four heads and six tails or five heads and five tails or four tails and six heads. The lowest number would be ten heads or ten tails.

Heads and tails coin distribution forming a bell-shaped curve

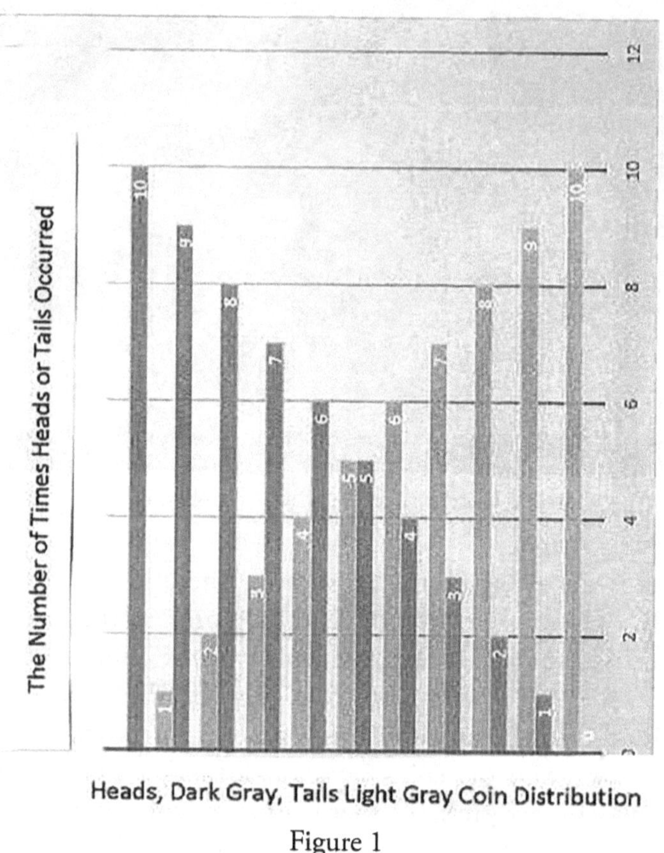

Figure 1

Figure 1 is a graph that demonstrates the combinations distribution of the ten pennies dropped on to a surface. The darker bars are heads, and the lighter-colored bars are tails. Looking from left to right, the bar farthest left are all ten heads. The bars to the right of the ten heads are nine heads and one tail. To the right of the nine heads of one tail bars are eight heads and two tails bars. To the right continues with seven heads and three tails. This is followed by six heads and four tails then five heads and five tails. This appears in the center of the graph with the bars for the number of heads and tails are equal. This continues with the remainder of figure 1 showing an increase in tails and a decrease in heads with the bar graph to the far right identifying ten tails. Where there are two bars together, identifying heads and tails appears as a bell shape where the top of the bell indicates five heads and five tails. One side of the bell identifies nine heads and one tail. The other side of the bell-shaped curve has nine tails and one head. The height of the bars indicates the number of times the distribution occurred. For example: The number of times one tail and nine heads occurred when the pennies were dropped on to a surface is less than the times six heads and four tails occurred. This distribution forms the bell-shape of the curve.

Again, you may ask, why is this important? It is important in that the distribution of head and tails forming the bell-shaped curve is considered a "normal distribution." Again, what does it mean, and why is important? Suppose I told you that I dropped ten pennies to a surface twenty-five times and counted ten heads every time. You should question the accuracy of my comments since the data above makes the possibility of twenty-five consecutive times of all heads or all tails virtually impossible.

This is one example of how science and mathematics destroys misinformation and conspiracy theorists' incorrect comments.

This bell-shaped curve is of great importance in statistics. For example: If I received the grades of one hundred students taking a test, where the grades that could be achieved were between 0 and 100 the test scores of the students would also resemble a bell shaped curve, with fewer students achieving grades between 90 and 100 than achieved a grade between 60 and 89. In addition, the number of students achieving a grade less than 60 would be less than the number of students that achieved a grade of 60 to 89; but somewhat similar to the number of students that achieved a grade between 90 and 100.

So, you may be saying, "All well and good, and that seems interesting, but I still do not get why it is important." All right, let us say that within the students above, there was one grade of one student that was a 2 on a scale of 1 to 100; and this grade of 2 is affecting the average grade of the entire one hundred students by pulling it below a 70, passing grade. This reflects negatively on both the students who took the test and the teacher. If we could find a way to not include the 2 grade, the average for the class will improve above the passing grade of 70. It is incorrect to just drop the grade of 2 from the calculation. A statistician could conduct a statistical test to determine if this 2 grade is real or statistically outside the range of expected grades. Meaning, something happened to the student that received a 2 grade, and the grade should not be included with the other ninety-nine student grades. It is not unusual to learn later, after investigating, because the statistician said the 2 number was outside the expected range of numbers, that the student was ill at the time of the exam and was not able to prepare for the test. Critically import-ant here is that with the findings of the statistician, the grade of 2 can be dropped from the calculation, and the class average is above 70. Without the statistician, the class average grade remains below 70 with, at best, a footnote that the 2 grade seems out of line but was included in the calculation.

I recently heard a discussion concerning the pluses and minuses of mail in voting and fraudulent elections. One of the commentators said that there were a series of studies done to determine if mail in voting will result in a fraudulent election. According to the commen-tator, all the studies found that there was no statistical significance identified for the claim that mail in voting will result in a fraudulent election. I do not want to address the politics associated with this issue. I want to address the science, specifically the words, "no statis-tical significance."

I shall begin by looking at the definition of the words, "statisti-cal significance."

The definition:

As a noun

1. The quality of being worthy of attention, importance.
2. The meaning to be found in words or events.
3. The extent to which a result deviates from that expected to arise simply from random variation or errors in sampling.

Statistical significance refers to the unlikelihood that a result is obtained by chance, i.e., it refers to the probably that a relationship exists between two variables or not. Critically important here is, we are talking about a statistical analysis being done to determine if, or if not, statistical significance exists. In other words, this is not one person's or a group of people's opinion. It is because of a detailed mathematical study designed to compare two items or variables, mail in voting and a fraudulent election to identify if the two variables have or have not a relationship with one another. In this case, the result was, there is no relationship between mail in voting and fraudulent elections, but there was a statistical significance between the grade of 2 and the grades of the other ninety-nine students.

I have chosen to spend some time on the collection and evaluation of data. I am doing this to provide examples of how COVID-19 statements that involve science require extreme rigor to ensure that the statements are accurate and correct. Too many people believe that statements from misinformation carry the same weight as statements from scientists. Nothing can be further from the truth.

Another example of the value of mathematical evaluation involve dice. Each die has six sides with dots from 1 through 6 dots on each surface of a six-sided die. When each of the two dies is rolled together on a surface, two of the die surfaces end up "faceup," with each die displaying several dots. The summation of the dots on the two die surfaces is then recorded. The recorded number can be 2, 3, 4, 5, 6, 7, 8, 9, 10, 11, and 12. Because the dots are equal to 1 through 6 on each independent die, when rolled together, the number 7 can be displayed with one die displaying either a 2, 3, 4, 5, or 6 so long as the other die displays a 5, 4, 3, 2, or 1 so long as two dies total seven dots. Therefore, a recorded number of 7 is the highest probability of a recorded number. The lowest probability of recorded numbers, 5%, is 2, one dot on each die' or 12, six dots on each die; 11, 5 and 6 dots on the dice; and 3, a 2 and a 1 on the dice. A 10, 9, 8, 5, and 4 on the dice, once they have settled, have 10% probability of occurring. The number 6 and 7 on the dice, once displayed faceup, have a 15% probability of occurrence. "So why is all this important, and why should I be interested?" Great question! With all this review, we can predict winning probabilities if one plays the casino game of craps.

12—6 and 6, 6 and 6, two probabilities, one roll, 5%

11—5 and 6 and 6 and 5, two probabilities, 1 roll, 5%

10—6 and 4, 4 and 6, 5 and 5 and 5 and 5, four probabilities, two rolls, 10%

9—5 and 4, 4 and 5, 3 and 6, 6 and 3, four probabilities, two rolls, 10%

8—2 and 6, 6 and 2, 4 and 4, and 4 and 4, four probabilities, two rolls, 10%

7—2 and 5, 5 and 2, 3 and 4, 4 and 3, 6 and 1, 1 and 6, six probabilities, three rolls, 15%

6—3 and 3, 3 and 3, 4 and 2, 2 and 4, 5 and 1, 1 and 5, six probabilities, three rolls, 15%

5—2 and 3, 3 and 2, 4 and 1, 1 and 4, four probabilities, two rolls, 10%

4—2 and 2, 2 and 2, 3 and 1, 1 and 3, four probabilities, two rolls, 10%

3—1 and 2, 2 and 1, two probabilities, one roll, 5%

2—1 and 1, 1 and 1, two probabilities, one roll, 5%

Therefore, for the game of craps, the probabilities of numbers displaying faceup after two die are rolled together are as follows:

The displayed numbers totaled	The % probability of occurrence
12	5%
11	5%
10	10%
9	10%
8	10%
7	15%
6	15%
5	10%
4	10%
3	5%
2	5%
	100%

The following line graphic, figure 2, depicts the order of things when it pertains to games involving dice.

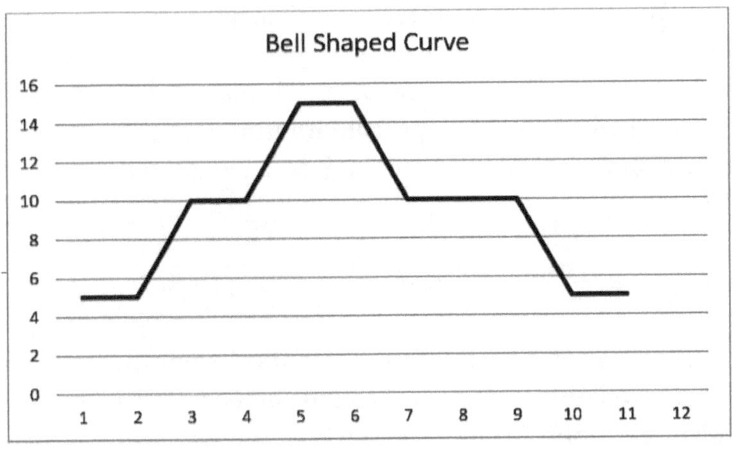

Figure 2

The Y axis displays probabilities of occurrence as a percentage. The X axis displays the total number when two die are added together.

- Numbers 5 through 15 identify % of occurrences on the Y axis.
- Numbers 1 through 11 identify the displayed numbers on two die, 2 through 12.

So why is this so important when considering communicating science? It is important because science recognizes and supports order. The earth makes one turn on its axis every twenty-four hours, and one turn around the sun every 365 days and six hours. The six hours is why we add one day every four years to our calendar, February 29, and we call it leap year. It is the way chosen to catch up with the six hours we are short in making a full revolution around the sun each year. Adding one day every four years permits us a catch up for the missing time. These are more examples of order, supported by science. Most people are not aware that, for exactness of time, we have a leap year.

With the game of craps, we are talking about probabilities, the probability of occurrence. From the above, you can see the probability of the number from 2 through 12 displaying with each throw of the dice. See figure 2. Therefore, the payout by the casino is higher when the 5% probability numbers are displayed than when 6 or 7 are displayed when the 15% probability numbers are displayed. You are three times more likely to roll a 6 or a 7 than you are to roll a 2 or a 12.

Correlation coefficient: What is it, and why is it important?

Earlier I addressed the term "statistical significance" and that there was no relation determined to exist between voting by mail and fraudulent elections. From time to time, you may find that two variables do have a relationship, and you may then choose to determine how best to determine how close is that relationship. A correlation coefficient analysis will help in that determination.

Have you ever said, "When this happens, that happens"? A relationship is implied, but not necessarily identified. If a relationship exists between two things that appear to be dependent on one another, how strong is that relationship, how meaningful is it, and how can we make use of this relationship? All good questions. Some things are obvious, like there is a relationship between the amount of gas you put in a vehicle and how many miles that vehicle will travel before the vehicle runs out of gas. The efficiency of the vehicle is another factor that will impact how far the vehicle will travel before it runs out of gas. There is also a relationship of hospitalizations from COVID-19 and the number of unvaccinated people.

A statistical analysis procedure called correlation coefficient clearly shows that the unvaccinated have a greater number of COVID-19 cases compared to those that are vaccinated. This is very important. Now it is no longer he says, she says. It is what the scientists, with support documentation, says versus what the conspiracy theories says with no documentation, and misinformation also with no documentation. It is the correlation coefficient analysis when interpreted by an expert clearly identifies a relationship or not; and in this case of unvaccinated, with COVID-19 versus the vaccinated with COVID-19, there is a much stronger relationship between the unvaccinated and COVID-19 than the vaccinated and Covid 19. Conspiracy theories and those providing misinformation do not use this type of analysis, making their claim that is unsubstantiated and dangerous. For example, one of the conspiracy theorists has said that if you get vaccinated, you become magnetized. You can put a key to your forehead, and it will stick because you are a magnet. This claim is totally laughable, unsubstantiated, and false. The problem is that people believe the claim with no evidence supporting it.

In other examples, there is an apparent relationship, but it does not appear to be as strong as the example about fuel in a vehicle and how far the vehicle will travel before it runs out of fuel. For example: the number of people on a pleasure fishing boat and the number of fish that will be caught, e.g., you would think that the more people on the boat the more fish will be caught. Or you may have two fishing boats, one red and the other blue. A claim can be made that the red boat catches more fish than the blue boat, so given the option, choose the red boat to fish. All the above can be evaluated, measured, using the correlation coefficient statistical technique, to determine relationships. This is another example of science looking for proof to draw an informed conclusion and communicating results minus jargon and in a manner that is understandable, the operative word being understandable.

The outcome of a statistical correlation analysis is interesting. A perfect relationship exists between two compared items when the calculated relationship equals the number 1. If no relationship exists, the calculated number will equal 0. Therefore, the closer the calculation comes to the number 1, the stronger the relationship. Equally important is that the calculated number can be a negative number. Yes, the calculated number can be below 0, no relationship. The number can reach negative one (-1). When this happens, it may be said that the relationship between the two items is the exact opposite of the expected relationship. It should be clear that statistical analysis is much more than arithmetic. It is a tool used by scientist to obtain truth.

With the above said, let us look at more meaningful correlations. It has been said that the use of masks will reduce the number of cases of COVID-19. If we were able to monitor the use of masks, and COVID-19, cases we could establish a correlation hopefully approaching 1.0. This is being experienced today with COVID-19 cases rising among those not vaccinated and not wearing a mask. Since the primary entry point of most viruses is the nose and the mouth, the wearing of masks blocks these entry points and prevents people who have COVID-19 and are wearing a mask from spreading the disease. The use of masks to prevent disease has taken place with the significant reduction of expected number of influenza cases in 2020. This occurred because of the wearing of masks to protect against getting and disseminating COVID-19 has protected the population from getting influenza. Influenza was the virus in the 1918 and 1919 worldwide pandemic that killed more than 50 million people.

Unfortunately, the issue of masks wearing has become political. Those rejecting masks have no documentation to support their case. Their position is, "I am not wearing a mask because it interferes with my freedom." People saying this do not understand nor do they want to understand that taking the vaccine and wearing a mask creates the freedom they seek—the freedom to return to the life they had prior to COVID-19.

Contact tracing is another strategy to limit the spread of the virus. This means if a person is diagnosed with COVID-19, they identify all the people with whom they have had personal contact in the last few days. These people may have been exposed to COVID-19 by being with the person that has been confirmed as having COVID-19. The approach requires volunteers to contact all the people who were in contact with the person that has been confirmed to have COVID-19. The idea is to have these people quarantine and as a result reduce possible spread of the virus.

This can become an impossible task. For example: if we have two thousand cases of COVID-19 per day, and the average contact for each case is fifty to one hundred people per day, this includes things such as going to the food store, the drugstore, and to work at the office. I shall take seventy-five as an average number of people contacted in a day.

If we assume two thousand cases per day and seventy-five contacts, seventy-five people need to be contacted. That is 150,000 contacts per day. If we have two hundred volunteers, each will need to make 750 contacts per day. That is about ninety-four calls per hour for eight continuous hours. It is not possible to do this every day of the week. In addition, if I assume 50 percent will not be available on the first try to reach them, that raises the number of calls, including call backs, to about 140 calls per hour, or less than thirty seconds of contact per call. This is the primary reason contact tracing was not done to any degree in the States.

Modeling

Models are used to predict the future. The results of modeling are only as good as the quality of the information being entered into the model, hence the expression, "Garbage in, garbage out." Models are mathematical tools where information is converted to numbers and entered into the model. The output of the model are numbers that are converted into information that is shared with stakeholders, and people who will use the information generated by the model. An example is the modeling of the infection rates of the COVID-19 pandemic. The model will predict expected changes in infection rates, if any. Since the information being entered into the model is ever changing, the results coming from of the model can be of questionable accuracy.

Summary

This chapter explains that science supports outcomes with mathematics. For example: statistical analysis can identify when results are outside what is expected. The results are then considered statistically significant. When two items have a relationship, wearing face masks and a reduction of influenza cases in 2020, that relationship can be confirmed. In other words, claims by an expert scientist can be confirmed mathematically. This is the exact opposite of misinformation, which has no documented statistical support and must be discarded.

CHAPTER 4

Are You and the World Prepared for a Global Pandemic?

Are you and the world prepared for a global pandemic? Having begun addressing science and mathematics, I want to now look at the management of a pandemic.

A crisis may be defined as an unplanned incident, accident, or event that, either initially or in a short period of time, expands to a level that involves deaths, injuries, loss of property, the involvement of many people, and high costs. Couple this with the pandemic, which is prevalent over the entire world, and you clearly have the COVID-19 pandemic that has generated multiple spin-off crises. What follows is the identification and brief description of several pandemics that have occurred over the last one-hundred-plus years.

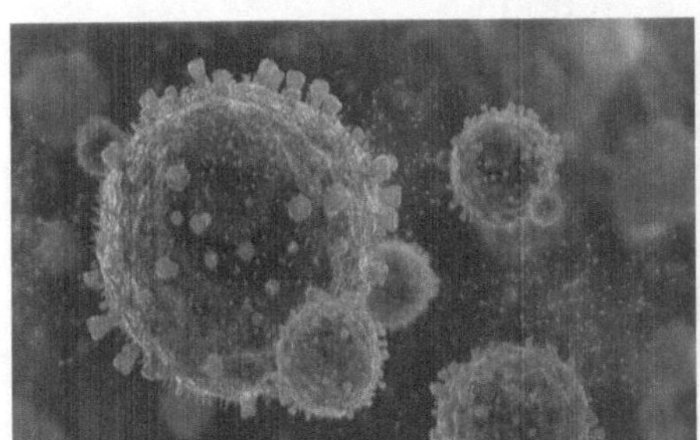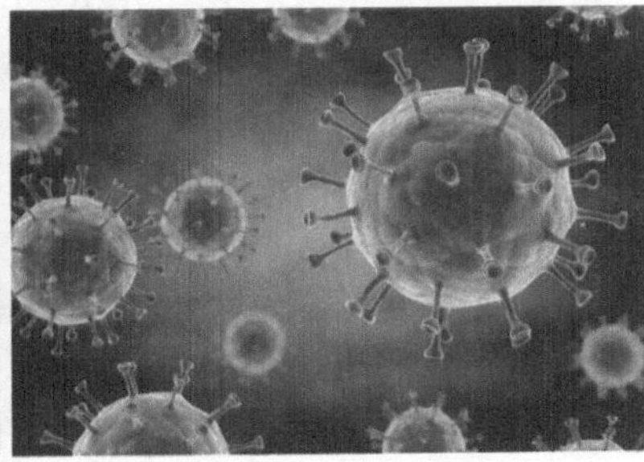

Viruses

Since 1918 the world has experienced three additional pandemics in 1957, 1968, and 2009. The subsequent pandemics were less severe and caused considerably lower mortality rates than the 1918 pandemic. The 1957 pandemic and the 1968 pandemic each resulted in estimated one million global deaths, where the 2009 pandemic resulted in fewer than three hundred thousand deaths in the first year. This decrease in severity of pandemic from 1918 through 2009 could very well have caused people to believe pandemics were fading away. The 2019 COVID-19 pandemic that has lasted into 2020 and 2021 and possibly 2022 and 2023 has taken away that thinking.

The world population has increased to almost 8,000,000,000 people. As human populations have risen, so have swine and poultry populations to feed them. This expanded the number of hosts provides increased opportunities for novel influenza viruses from birds and pigs and possibly bats to spread, evolve, and infect people. Global movement of people and goods also have increased to be an international plane flight away. Due to the mobility and expansion of human populations, even once exotic pathogens like Ebola, which previously affected only people living in remote villages of the African jungle, have managed to find their way into urban areas causing large outbreaks. This was highly visible with the COVID-19 virus.

Pandemics of the world

1. Plague of Justinian (541–542)
   Total deaths: twenty-five million
   Cause: bubonic plague
   This pandemic was thought to have killed perhaps half of the population of Europe. The outbreak afflicted the Byzantine Empire in the Mediterranean port cities, coming up to twenty-five million people in its year-long reign of terror.
2. The Black Plague (1346–1353)
   Total deaths: seventy-five to two hundred million
   Cause: bubonic plague

This pandemic ravaged Europe, Africa, and Asia. Thought to have originated in Asia, this pandemic most likely jumped continents via the fleas living on the rats that so frequently lived aboard merchant ships.

The physicians of the Middle Ages learned that many pandemic and other illnesses entered the body via the nose or mouth. We know this by observing head gear they wore during that time. See below.

Medieval physician costume designed to prevent
the physician from becoming ill

3. The third cholera pandemic (1852–1860)
    Total deaths: one million
    Cause: cholera
    Generally considered the deadliest of the seven cholera pandemics. It spread from the Ganges River Delta before tearing through Asia, Europe, North America, and Africa, ending the lives of over one million people.

4. The flu pandemic (1889–1890)
    Total deaths: one million
    Cause: influenza
    Originally, the Asian flu, or Russian flu as it was called, was thought to be an outbreak of the Influenza A virus subtype H2N2. The recent discoveries have instead found that it was caused by the Influenza A virus subtype H3N8.

Thought to be the first true epidemic in the era of bacteriology and much was learned from it.

5. Sixth cholera pandemic (1910–1911)
    Total deaths: eight hundred thousand plus
    Cause: cholera
    It originated in India before spreading to the middle east, North America, Eastern Europe, and Russia. American authorities having learned from the past quickly isolated the infected, and in the end, eleven deaths occurred in the United States.

6. The 1918 influenza pandemic
    This pandemic was the most severe pandemic the world experienced up until the 2019 coronavirus. It spread worldwide during the 1918–1919 time period. It was first identified in the United States military. The close proximity of troops led to a rapid spreading of the virus. It is estimated that about five hundred million people, or about one-third of the world's population, became infected with this virus. The number of deaths estimated at least 50,000,000 worldwide and about 675,000 in the United States.

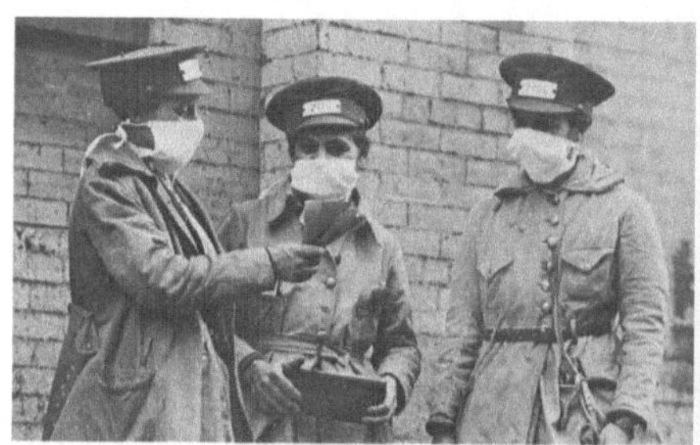

Military and the 1918 Influenza Pandemic

Pandemic US Impact

	Total Cases	Death	%	Decrease in Life Expectancy in years
1918 Influenza pandemic	500,000,000	50,000,000	10	12
2019 Coronavirus pandemic	254,288,809[1]	5,118,757[*]	2	1

Pandemic United States Impact

	Total Cases	Deaths	%
1918 Influenza pandemic	7,000,000(1)	675,000	10
2019 Coronavirus pandemic	46,697,360[*]	755,950	2

An estimate

The 1918–1919 data accuracy is somewhat questionable. Keep in mind that the world was at war. We know that troop movement was in large numbers and close quarters, a perfect combination for transference of the influenza virus, and many troops became ill and died. It was decided not to provide data on troop illness and death for enemy edification.

The 1918 influenza virus outbreak became known as the Spanish flu and lowered the average life expectancy in the United States more than twelve years. The name Spanish flu came about because reporting on the flu was blacked out in countries fighting in World War I. The only country reporting on the flu was Spain, so most people only assumed that Spain was the country where the flu began. This, of course, is incorrect. This disease spread around the world in the course of about six months, killing twenty-five million. Remember, vaccines were not available in 1918. The 1918 influenzas virus had a more devastating global impact than COVID-19 has to date. With no vaccine to protect against influenza infection and no antibiotics to treat secondary bacterial infections that can be associated with influenza infections, control efforts worldwide were limited to non-pharmaceutical interventions, such as isolation, quarantine, use of

---

[1]   November 15, 2021.

masks, good personal hygiene, use of disinfectants, and limitations of public gatherings, which were applied unevenly.

Mortality was high in people younger than five years old, twenty to forty years old, and sixty-five years and older. The high mortality in healthy people, including those in the twenty to forty-year age group, was a unique feature of this pandemic. While the 1918 virus has been synthesized, evaluating the properties that made it so devastating are not well understood.

The 1957–1958 pandemic

In February 1957 a new Influenza A (H2N2) virus emerged in east Asia, triggering a pandemic (Asian flu). This virus was comprised of three different genes that originated in the avian influenza virus. It was first reported in Singapore in February 1957, Hong Kong in April 1957, and in coastal cities in the United States in summer 1957. The estimated number of deaths was 1.1 million worldwide and 116,000 in the United States.

The 1968 pandemic

The 1968 pandemic was caused by an influenza (H3N2) virus comprised of two genes from an avian influenza, a virus. It was first noted in the United States in September 1968. The estimated number of deaths was one million worldwide and about one hundred thousand in the United States. Most excess deaths were in people sixty-five years and older. This virus continues to circulate worldwide as a seasonal Influenza A.

The 2009 pandemic

In the spring of 2009, another Influenza A (H1N1) virus emerged. It was detected first in the United States and spread quickly across the United States and the world. This new H1N1 virus contained a unique combination of influenza not previously identified in animals or people. Ten years later, work continues to better understand influenza, prevent disease, and prepare for the next pandemic.

The disease burden of this H1N1 virus continued from 2009 to 2018. This virus has circulated seasonally in the United States, causing significant illness hospitalizations and deaths. Additionally, the United States Center for Disease Control (CDC) estimates that 151,700 to 575,400 people worldwide died from H1N1 virus infections during the first year the virus circulated. Probably 80 percent of the H1N1 virus–related deaths were estimated to have occurred to people younger than sixty-five years of age. This differs greatly from typical seasonal influenza epidemics during which about 70 percent to 90 percent of deaths are estimated to occur in people sixty-five years old and older.

The 2009 flu pandemic primarily affected children and young and middle-aged adults. The impact of the H1N1 virus on the global population during the first year was less severe than that of previous pandemics. Estimates of pandemic influenza mortality range from 0.03 percent of the world's population during the 1968 pandemic to 1 percent to 3 percent of the world's population during the 1918 H1N1 pandemic. It is estimated that 0.001 percent to 0.007 percent of the world's population died of respiratory complications associated with H1N1 virus infection during the first twelve months the virus circulated.

The 2019 pandemic

The present pandemic began late 2019, continued through 2020, and with its present variants in 2021, is more than likely to be impacting lives in 2022 and possibility 2023.

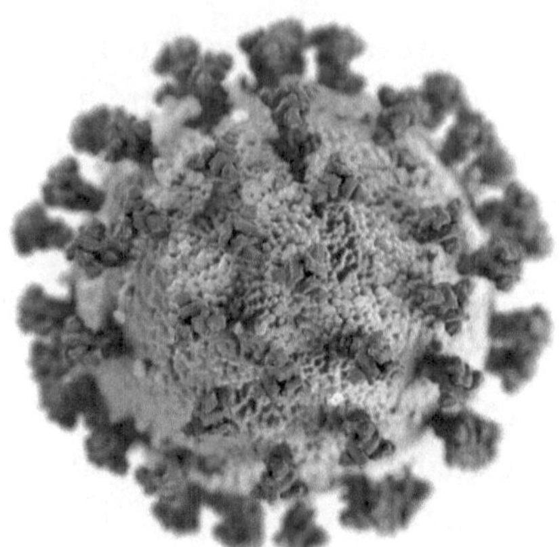

COVID-19 virus

The bottom line is that modern pandemics have been around for more than one hundred years, and they, or the threat of a pandemic, remains with us today. We are fortunate to live in an era where 90 percent of all the scientists that ever lived in the history of the world are living and working today. As a result, we have scientific breakthroughs such as vaccines now available to help us overcome the 2019 pandemic threat earlier than previously expected. While this is true, those who refuse to be vaccinated, along with the politics that support those who refuse to get vaccinated, are prolonging the pandemic, and people have and will continue to die unnecessarily. People have said that this COVID-19 vaccine to help fend off the virus has not had enough time in developing. If you consider the number of scientists that are living and working today, nine months for the development of a vaccine seems appropriate. In addition, virologists have been working with vaccines for more than fifty years.

Knowing the above, could you say with confidence that your leadership—government (all levels), para government or corporate—are fully prepared with programs procedures and tools in place to respond to a pandemic in an effective and efficient manner? How many would answer the question yes, and would you add, "Yes, of course," or would you say, "Yes, I think so. Yes, we should be able to," or "I'm not certain, but I believe someone is responsible to address this type of situation should it occur." In too many cases, it would be the latter comment. Again, what we know today about the COVID-19 virus, the answer is, "Yes, my leadership is prepared, and we are, can, and will address a pandemic if it occurs."

No one needs the programs, procedures, and tools to manage a pandemic until they are needed. Leadership needs to ask the question, "Is our pandemic management plan in place and effective?" And if the answer to that question is not a rapid and strong yes, much work such as testing the plan through a pandemic exercise needs to be done to ensure readiness. Money associated with the development, maintenance, and testing of a pandemic plan that may never be used, some may argue that the money may be more efficiently and effectively used elsewhere. To this comment, the question must be asked, what is the cost both financially and emotionally to millions of people dying in the world because of a pandemic? The answer is simple and straightforward: have a plan to address a pandemic, test the plan and make sure it works, and be ready to use it when it is needed.

Before an organization/country identifies or attempts to respond to a pandemic, the following concepts and questions should be understood, answered. and as appropriate, addressed:

- Are all unplanned events created equal, and if not, what are the criteria that distinguishes an emergency, crisis, epidemic, or pandemic from one another?
- What questions should leadership ask to determine if indeed a pandemic exists?
- How can a list of pandemic response resources be kept up to date?
- Is a pandemic team approach better than an individual or individuals addressing the pandemic, and if so, why?
- How many people should be on a pandemic management team, and why?
- If a team approach is chosen, who should lead the team, and why?
- How quickly should the pandemic team be gathered, and what happens if team members such as the leader is out of the country?

- Should there be a plan facilitator, and if so, who should it be, and what does he or she do?
- Should there be a pandemic management plan? If yes, who should write it, who should be responsible for it, who should approve it, and where should it be kept?
- With the pandemic management team in place, what is next?
- Should a pandemic management plan be a controlled document with an owner, and what is a controlled document?
- Who should have copies of the pandemic management plan?
- If each country has its own pandemic management plan, how might they be coordinated, and who should take a leadership role, i.e., the world health organization?
- When is the work of the pandemic management team done?
- Who should have copies of the pandemic management plan?
- What are the nine stages of a pandemic management plan?
- Should there be a pandemic plan manager, and if yes, what is his or her role?
- How does the speed of a pandemic, spreading throughout the world, influence leadership decision-making?
- If there is a need for outside expertise in a pandemic, who should that be, and how and when should they be contacted?
- How should the media be managed during a pandemic, and by whom?
- Who should prepare messages for the media, should they be approved, and how should they be provided to the media?
- Is media training a value?
- What happens when a pandemic reaches stability, and what is stability?
- How will crises resulting from a pandemic, e.g., loss of jobs, emotional impacts, and school closures, to name a

few, be managed such that they don't interfere with deci-sion-making to address the pandemic?

- Who evaluates the impact of the pandemic, and when?
- When should remediation and recovery begin?
- What is the value of a pandemic scenario?
- Should the pandemic management team undergo training?
- When should a pandemic exercise or exercises be done, who should conduct them, and how should they be carried out?
- What should be done with the result of a pandemic exercise?

A pandemic represents a global crisis. It is important to real-ize that too often the pandemic crisis will trigger additional crises. Many times, these additional crises appear more impacting than the pandemic. While the following will be addressed in detail later, it is important to identify the spinoff or sub-crises of a pandemic crisis and how to deal with them. Examples of a pandemic sub-crises include among others:

- Lockdowns of communities resulting in businesses closing and people losing jobs.
  - o I wish it was that simple. The stress and mental impact of not seeing loved ones and no money to pay mort-gages, car loans, tuition, among others, provides a feel-ing of helpless, distress, despondence, and depression.
- With no jobs, there will be rent or mortgage crisis, a finan-cial crisis.
  - o Having a roof over our heads is fundamental to life. A loss of a home causes people to look at their self-worth. This is not a good situation.
- Since we are very much a global economy, a factory in China shutting down will negatively impact automobile production in the states if they make a critical part needed to build a United States manufactured vehicle.
  - o Many people have already experienced this situation. Lumber has become extremely expensive. This will

raise the prices of homes. One of my children ordered a sofa for a home and was told she would need to wait three months for delivery and not the standard three days for delivery. Far more serious is a wait for needed medication. This is not helped with the announcement that mail delivery will slow in the United States.

Lack of Income

- Scarcity of parts for an automobile will also result in an increase in price of that part should they be available in limited quantity. The increase in the cost of the part or parts may increase the price of the vehicle that cannot be purchased because people that have been laid off or their businesses have been closed because of the pandemic do not have the money to purchase the automobile.
  o Similarly waiting for parts because factories are closed all over the world, but also, parts and other items and services are not available because some people have chosen to not return to work because they are receiving federal funds designed to help them through difficult times.
- This is to say nothing of the emotional impact on people not able to be with other people because of quarantine.
  o The emotional impact of the pandemic has not fully been evaluated. It is expected to be significant.

- Children loosing social contact with peers and teachers are expected to be severely socially impacted along with our educational systems. While online learning helped to some degree, many children do not have computers or internet and, as a result, missed a year of education.
  o Children are expected to be impacted by a lost year at school for years to come. For example: math programs build on year-to-year studies. Understanding fractions is very important to the understanding of decimals. Science also builds year to year. Missing a year of science will affect the following year of science curriculum.
- There are more spin-offs or sub-crises than those identified here, but for the present, the above sub-crises will be addressed.
  o Sub-crises are very important to understand and address as we make our way through this pandemic.

The importance of satisfying these spin-off or sub-crises cannot be understated; however, it must be remembered that in satisfying the sub-crises, the pandemic will not go away. However, if the pandemic is satisfied, the sub-crises will be satisfied quicker and easier than if the sub-crises are addressed and the pandemic remains.

A sub-crisis

A sub-crisis is a spinoff crisis from a pandemic. They include, but are not limited to:

- Business closures
- Loss of job
- Loss of income
- Loss of home
- Loss of car
- Lack of food

- Emotional issues due to impact of pandemic including the death of loved ones
- Supply chain issues due to international impact of pandemic
- School closures
- Remote learning does not work for students without computers
- Not enough staff to fill positions as businesses open

CHAPTER 5

How Do You Review the Risk and Prevent a Pandemic?

In a book about a pandemic and spin-off crisis management, it should not be surprising that it includes a discussion and an emphasis on risk. Simply stated, if hazards can be eliminated or risks reduced, the probability of crisis or pandemic will also be reduced. This includes many of the pandemic situations that occurred during the twentieth century and the beginning of the twenty-first century. It was in that time that the world experienced some of the worst pandemics, some of which have been identified and described in chapter 4. We have learned a great deal from past pandemics and, as a result, are able to employ strategies and corrective actions to ensure they should not occur again. What is known about the topic of risk? There is confusion between the words risk and hazard. A risk is not a hazard, and a hazard is not a risk. The COVID-19 virus is the hazard, and the risk occurs when you are walking toward a location where the virus is present. Another way to explain it is the following:

Children are walking across a field with tall grass, and fifty yards ahead of them, there is a hole in the ground. If not careful, one or more of them could step into the hole, that is not visible because of tall grass, and sprain his or her ankle or experience an even worse injury. As the children approach the hole, their risk of injury increases because they are drawing closer to the hazard. If they make a slight turn to the left or right as they proceed across the field, their risk is reduced because they are no longer walking directly toward the hole. If they turn back toward the hole, their risk increases. If the children turn around and move away from the hole, their risk of falling into the hole no longer exists.

If a person is not vaccinated and enters a location where the virus exists, there is a high probability that the hazard, the virus, will enter the person and infect them. The use of a mask is another barrier to prevent the virus from entering one or more of the two key entry points for the virus, the nose and the mouth. Without this mask barrier and not having been vaccinated, there is a very high probability the person will come down with the virus.

Risk hazard—means of transport—*receptor*

While risk and hazard are related, they are different. Risk has been defined as the probability of damage, injury, liability, loss, or other negative occurrence caused by external or internal vulnerabilities and that may be neutralized through meditated action. A hazard is anything or anyone that can cause harm. To have a risk, one must have a source of a hazard, a receptor of the hazard, and a means of transport from the hazard to the receptor or from the receptor to the hazard. Remember, to have a risk, you need a hazard, the hole in the ground that children cannot see; a receptor, the children; and the means for transport, where in this case, the children are walking in the direction of the hazard. The hazard, the virus, present in a room, the person is the receptor of the virus, and the means of transport is the walking into the room in the building where the virus exists and may impact the receptor.

Everything that we do has an element of risk, and we can and do live with risk within acceptable limits; however, the trick is to recognize risk when risk moves outside the acceptable limits. No part of a pandemic is within acceptable limits. A pandemic or spin-off crisis situation may occur when a risk moves outside of acceptable limits, and that movement is not always recognized. How many times have you found yourself in a conversation while driving a car, and when you looked down at the odometer, you were exceeding the speed limit? You have experienced an increase in risk without immediately recognizing it.

There is a vocabulary associated with risk. For example:

Risk identification—this involves determining whether a risk is real or is perceived. Perceived risk can be as time-consuming and concerning as a real risk.

Fear—fear has been defined as future fantasy expectations appearing real.

Real risk. For example: skiing, skydiving, boating, especially without the ability to swim, bungee jumping, flying in an airplane, taking a train, and driving a car all involve risk; but the risk may be judged acceptable. Risks are acceptable to the people who engage in such activities since they have made the determination that what they are about to do is an activity that they have determined they will not be killed or injured; and therefore, the risk is acceptable. Unacceptable risks are usually those that are forced upon an individual. When this occurs, the individual often fears for their safety, the safety of their family, the safety of personal property, the safety of others, and the safety of the environment. When risks are judged to be acceptable, there is a feeling, *I am in control of the situation.* Conversely, risks are judged to be unacceptable when a person does not feel they are in control of the situation. There are times when people and their loved ones are faced with unacceptable risks because of decisions they were not aware of or decisions where they may have had no or minimal input, for example, decisions of local or state government transportation departments as to where to put a stop sign or a traffic light.

Government transportation departments receive funds every year to make changes in road configuration and the placement of traffic lights and stop signs and other traffic control devices. Decisions are also made regarding the placement of overpasses in congested areas. Transportation departments are not given enough money to make changes to every street corner or area where there are potential risks.

A site is selected based on criteria and information received; however, often this decision is made without public input, or input from a public hearing. As a result, at one intersection, the risk for accident and injury remains the same, while at another, the one where the new overpass is constructed or where a traffic light is installed, the risk has been reduced. It is important to point out that the transportation departments are not at fault; they have a limited budget and limited criteria from which they make informed decisions.

There are many different types of risk. There are personal risks as they relate to self and family. There are financial risks. These are risks to the economy and the public economic well-being. This often results from a pandemic. There are risks with the potential to impact community financial situations due to loss of jobs because of shutdowns and closures because of a pandemic, and there can be multiple risks from multiple hazards. These have the potential for multiple impacts such as the potential to impact people, personal property, public health, the economy, and the environment. Operating a small boat alone with an outboard motor is only one example of multiple potential hazards that may lead to risks. There are potential hazards such as weather, the functioning of the outboard motor, and the condition of the boat. The possible risks are to your well-being should a severe storm develop, the motor overheating and stopping, and the risk that the boat will leak and possibly sink. Having said this, judgments are made before getting into and operating a boat that all these potential hazards and possible risks are acceptable.

To be effective in reducing or eliminating a risk, the risk must be measured. For example: normal body temperature is 37°C or 98.6°F. Regarding blood pressure, systolic measurement of 120 and a diastolic measurement of 80 in the United States are generally considered to be normal. Individuals with above-normal blood pressure may also be considered at risk. Body fat and body weight can be measured. Both can identify a potential risk. Gene mapping will identify personal hereditary health risks. Measurement of risk is not limited to health. Your height above sea level while standing on the side of a mountain, statistical measurements of illness in each community, the purchase of stock in a weak company, the driving on tires with little to no tread, and the possibility of being struck by lightning are examples of measurable risk.

The decision on the part of the local transportation
company regarding where to locate a traffic signal

What do mathematical models do, and why are they important? Mathematical models predict the future. The better the input data for the model, the greater is the accuracy of the prediction. Accounting for variables is the problem in the accuracy of model predictions. Today most people are familiar with the models used to predict weather. While the weatherman or lady tend to be correct some 80–90 percent of the time, we remember most the picnic that was rained out when it was not supposed to rain. Therefore, models are not 100 percent accurate. Small changes in barometric pressure or the development of an unexpected low-pressure area often results in unpredicted weather. Models are certainly not perfect; they are very important today in that they can help people prepare for the possibility of hurricanes and tornadoes.

While it is critically important to identify risk and put in place that which will reduce or eliminate risk, it is equally important to prevent risk, which may prevent a pandemic or spin-off crisis. The ability to predict risk using models is also a part of our culture. I said, the weatherman's/woman's use of mathematical models tells us that rain is coming, a hurricane or a tornado is coming; government health departments tell us that the flu season is upon us; and downstream impacts because of pollution in the breathable air and water used for recreation and if water conditions are right for drinking. There are also high-level risk models that are used today. These models include health risk models, catastrophic risk models, financial risk models, credit risk models, stock market risk models, and hurricane prediction models, among others.

There are some simplistic predictive exposure models for risk such as the one that follows. This model has been used in attempting to predict community exposure to chemicals in the ambient air or breathable air. Using 168 hours in a week and considering 40 of those hours as a workweek, a risk factor may be created. This is done by dividing 168 by 40 to get a factor of 4.2. The idea behind this is that while at work at a chemical plant, a person is more likely to be exposed to chemicals than a person at home or working in an office. Using a typical workweek and considering the number of hours away from the exposure to chemicals and assuming the person is healthy and has not gotten ill from exposure to a chemical to the worker, the level of exposure of the worker appears to be acceptable. Obviously, there is little or no science applied to this simplistic model. The thinking is to apply this logic to people who do not have exposure to chemicals.

Household chemicals

Exposure to chemicals can be grouped into three areas (1) exposure to chemicals with a potential to cause minimal to no harm to people, (2) exposure to chemicals that may cause some harm to people, and (3) exposure to chemicals that may cause harm such as cancer if people were exposed to them. Obviously, this approach does not take into consideration the general health of the exposed population, the genetic makeup, or susceptibility due to heredity of the exposed population, nor does it address concentration and time of exposure. With that said, the following approach may be used to get a general understanding of the possible risk of exposure to air pollutants.

Considering a lethal dose (LD50), the 4.2 factor to identify possible risk may be used. The LD50 is the lethal dose concentration level of exposure to a chemical where 50 percent of an exposed population is expected to die. This is typically a very low concentration level to protect people and is often generated from animal studies. The number is typically used in factories where exposure to chemicals is measured and documented. The following is an example of how to use this simple predictive model described above to identify possible human exposure to ambient or breathable air concentration levels of chemicals.

For the chemicals thought not to cause harm from exposure, the LD50 is divided by 4.2 to get an ambient breathable concentration level. For chemicals thought to cause harm from exposure, the LD50 is divided by 42. For chemicals thought to cause serious harm, such as cancer, because of exposure, the LD50 is divided by 420. As an example, for chemicals considered low risk, having an LD50 of 0.1 part per million (PPM) divided by 4.2 will represent an ambient or breathable air exposure level of 0.024 ppm. Chemicals of moderate risk with an LD 50 of 0.01 ppm are divided by 42, which would result in an ambient air concentration level of 0.00024 ppm. Chemicals of high risk (cancer-causing) and with an LD 50 of 0.001 ppm divided by 420 would produce an acceptable ambient concentration level of 0.0000024 ppm.

Please keep in mind that the above approach is no guarantee that there will or will not be an impact from a risk. If models are used, they should be considered only an indication of possible impact or outcome from a risk. As a rule of thumb, if the predictive model described above used in the chemical exposure number is above acceptable ambient concentration number, a more-sophisticated model should be used, and possible testing should be done to determine the actual community exposure concentration.

This analysis is presented to provide only a small example of scientific studies that are done to protect populations from exposure to chemicals. Since COVID-19 is a complex chemical protein, this type of analysis could be used. However, there is no LD 50 for the COVID-19 virus. Any concentration is not considered acceptable.

Crisis prevention

Over the past twenty or so years, the world has experienced several crisis situations, including pandemics. Whether they have been man-made or an act of nature, they are often devastating, resulting in loss of life and property. The question is, how can such incidents or events be prevented? Further, as a world community, have we become accepting of the idea that events will occur that all too often result in a pandemic or crisis? What steps can be taken, and what can be put in place to minimize the possibility of events that may lead to a pandemic or spinoff crises? As previously stated, a hazard is not a risk. It is a potential source of harm or impact where deviations from design or operating intent may occur, resulting in a risk. Harm may be the consequence of a hazard and may result in death, physical injury, damage to property, the economy, or the environment.

Information is critical in order prevent incidents, events, and situations that may lead to a pandemic or the many crises that are spinoffs from a pandemic, e.g., shutdowns impacting the economy, job loss, mortgage foreclosures, and the like. Those who have knowledge and experience regarding managing a pandemic or crisis and are familiar with the situations that may lead to a pandemic or crisis are invaluable. For example: if you enter a hospital with chest pain, does this represent a crisis? An electrocardiogram (EKG) is taken, and blood is drawn to test for enzyme levels. Trained people are called in to test and evaluate your condition. These include the medical technologists who will draw the blood, technicians who will conduct an EKG test, and a physician is called in to read the results and to examine you. "It is the knowledge and experience of the physician and those who drew your blood and conducted the tests along with the test results that the physician uses to provide you the information as to whether or not you have had a heart attack." If you have not had a heart attack, you go home happy—there is no crisis. If you have had a heart attack, you are frightened and fear for your life; for you, this is a crisis. Planning and commitment are required to prevent a crisis or a pandemic. Time-tested techniques such as quarantine, social distancing, the wearing of masks, do not gather indoors with large groups, get vaccinated if a vaccine is available, contact tracing, and many others should be employed to prevent the spread of the disease.

CHAPTER 6

Communications during a Pandemic

When it comes to pandemics and spin-off crises caused by a pandemic, efficient and effective communication is critical. While this seems simple, it is not because not everyone has the same agenda, are not on the same page, nor do they have the same target audience. We must effectively communicate and should all be on the same page to stop a pandemic and return to a "normal" life.

As with any natural disaster or upset condition including the COVID-19 pandemic, the concerns must be with those that have been infected and their families. While the impact may affect jobs, the first concern is the impact on the lives of the members of the community experiencing the pandemic and thereafter spin-off crises. Specific concerns must be addressed for those identified as critical receptors. Generally, these are the aged, the young, and the infirmed. History tells us that these are the people who are most impacted during a pandemic. The aged because many are already suffering from illness, or preexisting condition; the young because the impacts could affect the rest of their lives; and the infirmed because these people are ill, in need of care, and are susceptible to a pandemic. If someone who is infirmed, aged, or both is displaced during a time of a pandemic, it may be critical that they have their medication or access to their medication should they be evacuated from their homes or from wherever they reside.

Hospitals must be prepared to receive patients experiencing respiratory issues typical of the COVID-19 pandemic. A caller from the hospital to local government official should be put in touch with an industrial hygienist or physician. Communications between hospital, patients, and loved ones has been difficult if at all. Communities are usually prepared for events such as tornadoes and hurricanes, especially if there is a history of such events. A pandemic is a different situation. The pandemic virus is a colorless, odorless, and tasteless material that can have devastating impacts.

Skill sets such as organizational, interpersonal, and leadership must be in place and working when upset conditions exist that may impact a community. At this time, leadership must be clear and visible in the community. The two, leadership and visibility, must be seen as occurring to minimize the impact and maintain control of the situation. The communications which, if possible, should be joint, must come from community leadership (the mayor, police chief, among others) and the scientific leadership. Communications must be joint, specific, and done in a timely manner. Each time messages are delivered to the community. questions should be taken from the media. The responses must be clear and to the point. Messages that are important for the community to hear may not only be presented at the outset of a press conference but also should be restated, where possible, as part of responses to questions.

The individual who responds to the media must be trained on how to speak to the media, must be empathetic and sympathetic to those who have been struck ill and the families of those who have died because of the pandemic. Community leadership and the scientific community leadership must speak in the same voice to obtain the confidence of all involved. Since those in charge are expected to know the answers to questions, this could be a problem. If the person from the science community is asked to answer a question and they do not know the answer, this non answer, or incorrect answer, will cause a loss in credibility. It is for this reason it is best to have a trained public affairs person address the media. The public affairs person is not expected to know the answer. Therefore, when the public affairs person says, "Great question. Shortly I shall return to the team managing this pandemic. I shall get the answer and share it with you in a couple of hours when I return for the next briefing."

Unfortunately, this is not occurring in most places within the United States. If this is not done, there will be information gaps and interdependencies that are not clear. The community and scientific leadership working through their public relations people must not stonewall a response in any way. A problem occurs when leadership and the science are not on the same page. Science has the message that we need to be vigilant, maintain a social distance of six feet, get vaccinated, wear masks, and much more; and leadership is pushing the message not to worry, this will all disappear. This is a classic scenario what typically happens when leadership does not like the message the scientist is presenting. The simple answer to the problem is shoot the messenger, the scientist. Not literally but discrediting the messenger will do just as well. While this can be done, it does not change the message. However, it takes resources away from the effort to address the virus. It focuses on what can we say about the expert that will lower his or her creditability and allows the public to focus on the claims of the leader. The media will soon carry stories that the scientist said one thing then change their mind. The implication is, they do not know what they are doing, when in fact the virus mutated, which changed the recommended path forward.

The fact that the virus mutated is not mentioned in the media article and TV comments. Since the message of a need for more vigilance, vaccines, mask-wearing, and the like is still front and center, and no matter what the leader says, the virus cannot be wished away, it will return with a vengeance should the community not follow the advice of the scientists. The affects can be devastating in terms of hospitalizations and loss of life. Leaders must understand that science is not a person. By discrediting the person that talks about the science does not affect the science. Indeed, it is an attempt to discredit science by discrediting the messenger. Discrediting science is difficult because of the enormous amount of documentation that supports the science. Therefore, there is too much documentation supporting the science to discredit it, and there is no documented support that the virus will just disappear. I recently heard an individual make a claim about a past event, where he said the event was controlled by an outside group that influenced the result. When questioned about documentation supporting his claim, he said, "Why do you need that? Just report what I am saying." Unfortunately, many of the followers of the leader will support whatever he or she says, even though there is no documentation supporting the leader's words and a huge amount of science documentation supporting what is being said by the science community.

People are tired of being confined by the mandates of a pandemic, and they want it to be over. There is a way that it can be over: get vaccinated, wear a mask, wash your hands, maintain social distancing among other documented approaches to protect yourself from the COVID-19 pandemic. People are not willing to do what is required. They say it affects their freedoms. Their actions will prolong the pandemic.

Science communication

Communicating science is often a challenge. Most scientists who have spent their life in some area of science have a great difficulty communicating their knowledge, not necessarily in the classroom but to the public. In the classroom, the scientist has time to bring the class along and make them acquainted and comfortable with jargon. When speaking to the public via the news media, the scientist must deliver a one-sentence explanation and has no time to explain the science behind the comment. The explanation of scientific thinking must be clear and relatable. Jargon must be replaced with meaningful words that make sense to the reader or listener. For example, a text with jargon would read, "The chemical was found to be ten to the power of six times more potent than originally expected." The same text without jargon would read, "The chemical was one million times more potent than originally expected." While this may appear clearer to the reader or listener and straightforward without jargon, the scientist must spend hours to find the correct words for clarity.

Remember, while all the public wants to hear is all is well and they should not need to worry about their family and especially their children, on the other hand, the scientist does not want to say something that may be misinterpreted or not correct. The problem is variables. Nothing is exact, but the scientist worries that their words will be taken as the exact solution to the issue or problem.

Since there is always a possibility, although slight, that what the scientist is saying in interpreting data is not 100 percent accurate, the scientist will often hedge in saying that the results are 100 percent accurate. The public wants to hear "good news" without the hedging. For example: wearing a mask in public will help to reduce exposure to the COVID-19 virus. The public wants to hear that if they wear a mask, they are 100 percent protected from the virus. The variables that prevent saying that masks are 100 percent certain to prevent infection from the virus include the construction of the mask, the closeness of the mask to the face, if the mask has a 100 percent seal between the skin and the face mask. Is the mask cleaned regularly? If the mask is not cloth, is it used only once and thrown away? And many more.

When I taught for a brief period, I would tell my class, "You see me in front of this class, you may judge that I know this subject because if I did not, I would not be here to teach you. My issue is not to overwhelm you with this subject nor is it to demonstrate my knowledge. It is my job to find you and provide you information about this topic such that you can absorb it and it becomes meaningful to you. If I cannot do that, it means that I have not found you as a point in space, and I have failed to communicate." This is the same as communicating science to the public. If the science community cannot effectively communicate to the public, it has failed.

Politicizing science

Politicizing science is wrong and dangerous. Science is not a person that can be blamed nor criticized. We trust people who are trained to explain science. They are usually multi-degreed with years of experience. By putting pressure on medical experts, public health agencies, and the like to move outside of protocols that have worked for years to say things that are not correct is dangerous and causes a loss of creditability. Leadership must understand that science cannot and will not be put under pressure to be steered in a desired direction. Science will move in the direction of truth in its own time. It must not be rushed to incorrect conclusions that will put countless people at risk.

Regarding COVID-19 treatments for those who have the disease, many people have encouraged unproven treatments, saying, "What do you have to lose?" There is a great deal to be lost, including people's lives. Treatments are designed to build the body's immune system and help the body to recover. Treatments do not cure the disease. Treatments are designed to have the body overcome the virus. I have tried to come up with a way to explain what it means to rush a vaccine or a treatment. I shall liken it to the building a modern vehicle with all its components. What if, at the one-half or two-third point of construction of your vehicle, it is decided to rush the completion of the vehicle because you would like it in your hands? There are many questions. The obvious question is, "Is the vehicle safe to drive?" A salesperson is telling you, "Of course, it is safe to drive." Is it going to satisfy you once it is in your hands and in light that it was rushed to get it to you? You want documentation that proves it is safe. You would ask questions concerning quality control, and were the breaks tested before you received the vehicle. This is no different than being in receipt of the vaccine. You would expect that the vaccine would be fully tested before you receive it. You would not want to hear from a politician saying that they had every confidence that this "pushed through" vaccine is safe. You would want to hear from the head of the health agency and/or from someone intimately familiar with vaccine-testing protocol who will assure you that all the required testing for the vaccine is completed and the vaccine has been approved for use.

Recently there has been a flood of misinformation concerning masks and vaccines. One of the problems is that the politicians want you to believe that everything they are saying is just as creditable as what the scientific community is publishing. There is nothing further from the truth. The politicians say what they say with virtually no documentation while the scientific community cannot publish anything without rigorous study and documentation. Therefore, the comments from the scientists are much more accurate and correct. However, no one in the scientific community steps forward and says loud and clear that the statements of the political community are wrong. No more than twenty-four hours should pass, and the misinformation should be trashed with strong documented scientific evidence or request scientific evidence that supports the misinformation. If this is not done, and the misinformation is allowed to stand unchallenged and it is taken as factual by the larger community, it takes on a life of its own.

Misinformation creates fear. It is that simple. Fear is a tool used by conspiracy theorist to place doubt in the minds of people. That doubt creates delays in the effort to rid the world of COVID-19. Fear has been defined as future fantasy expectations appearing real. While the definition gives one pause, it does not make the fear disappear. Undocumented misinformation such as, "If you receive the vaccine, you will become magnetized," has resulted in people placing metal near their body to see if it would stick. As difficult as that sounds, it happened.

No one wants to be afraid. Typically, people run from or move away from the issues or items causing fear. The conspiracy theorists have been successful in spreading misinformation and caused people to move away from the vaccine, resulting in prolonging the pandemic, and as a result, more people have been hospitalized and are dying. Looking at the success of the vaccine in immunizing people and preventing major impacts from COVID-19 should make more people want to get vaccinated. There is an expression, "The truth will set you free." Unfortunately, people do not know and are seeking truth. They will take the misinformation over science because they do not trust the science. Science needs to do a better job in communicating truth. Please remember, when a scientist speaks about data, they are providing an interpretation, not an opinion. There is a huge difference between the two. Interpretations are comments based in education and experience. A conspiracy theorist provides opinion without documentation.

What the public wants to hear: the community and a pandemic or crisis

Community fire, police, and ambulance services are typically well trained and knowledgeable in addressing emergency, crises, and pandemic situations. They need to establish a command center to address the situation. When this occurs, communities that get involved in the emergency or pandemic situation are responsible to supply needed information to the command center. Remember, at a time of emergency, crisis, or pandemic situation that impacts a community, there will be stages and, if possible, information providing assurances to the community that the government and local services working together will minimize impact and bring this event to stability as soon as possible. There are legal issues associated with any event that involves the community. While every effort is being made to save lives, minimize impact, and bring the event to stability as quickly as possible, you must be careful to stay with the facts. Do not offer opinion even when asked. Unfortunately, many involved outside the scientific community provide opinion and undocumented information, and when asked for the documentation, they respond with why do you need that. "Believe me, and do not worry about proof. It is not needed. Just believe me."

Who or what is responsible for the pandemic is not a subject for discussion, especially during this time of minimizing the impact and bringing about stability! If asked how this could happen, the recommended response is, "I do not know the answer to that question, and I am sure it will be addressed later. For now, our concentrated effort is on saving lives first, aiding the ill, and supporting the families that have been impacted." At no time place blame or take responsibility for a pandemic.

As with any pandemic, there is great potential for people to die. There is fear on the part of the families for the aged, the young, and the infirmed. No one wants to tell a family that their loved one is alive and well, only to follow up some time later that their loved one has died because of COVID-19. In most cases, when informing a family of the loss of a loved one, it should be done in the presence of a medical doctor, a member of a professional counseling team, and perhaps clergy who can immediately begin working with the family. Similarly, lifetime friends' coworkers who have worked together for twenty or so years are like family. A loss of a coworker or lifetime friend may be as traumatic as a loss of a family member. Therefore, leadership must be sympathetic and empathic to the extended family as well. The involvement of counseling for coworkers is also not out of the question. While leadership needs to deliver messages as to progress being made to bring the pandemic to stability, their messages to the larger community must reflect the loss of those who have died in meaningful, sincere, sympathetic, and empathetic terms.

Why is it so difficult for science to be accepted by community? To begin, science is difficult. While high school curriculums, for those who look to possibly attend college, generally try to include general sciences in the first year, biology the second year, chemistry the third year, and physics in the fourth year. I also want to make sure to include mathematics in this grouping. While this may vary from school district to school district, many students try to avoid taking science courses, saying, "I do not need it. I shall never use it and/or it is too difficult." As a result, people who follow through with college, graduate school, completing bachelors, masters, and PhD science degrees, are few as compared to the total population. While the reasons for not taking science courses are similar throughout the world, the world outside of the United States appears to be far more successful having students taking science and mathematics than the United States. The result is more international students taking graduate science courses and degree programs in the United States, leading to masters and PhD degrees. The irony here is that the United States has many science students coming from other countries. This clearly places United States science graduate degree programs in high demand globally.

A science and mathematics curriculum builds from topic to topic. For example: elements of biochemistry that addresses items like how the body takes food and creates energy need curriculums in chemistry, biology, nutrition, and food science to be understood. Therefore, a weave of multiple scientific disciplines are necessary for the creation of new thinking and new products, e.g., vaccines.

Science textbooks

So why is something so important rejected by the public? If a notion is presented and is thought to be too difficult to understand, it is much easier to reject the notion than accept it. This rejection limits the embarrassment of saying, "I do not understand the teacher." Often, if there is support for the rejection, even though there is no documentation supporting the rejection, people feel more comfortable with their rejection pointing to a person they respect who has rejected the notion even though there is no support for their rejection, e.g., the rejection of the COVID-19 vaccine.

This is a failure in two areas, those in the science community who do not have public trust and those in the community who reject the notion because they do not understand it and do not wish to take the time to learn more, again fearing the embarrassment of not understanding, that may convince them of the value of the rejected notion. Let us begin with the lack of trust of the science community. Many people do not know why people not in medicine are called doctor. The title of PhD literally means philosophy doctor. It is the highest degree awarded throughout the world in each discipline, whether it be science, social science, the humanities, or other. In many areas, its importance is considered a priesthood, or in parts of Europe a professorship at a university. Medicine does not offer a PhD degree.

Generally, people are intimidated by doctors. In a physician's office, people often do not understand what the doctor is telling them but will not question for fear of appearing stupid, or are given medication but will not ask why the medication is necessary, or if there is another approach, without the medication. The trust level for a physician is generally higher than a PhD because there are far more one-on-one visits with a physician than a person with a PhD in science. Most people with PhD degrees do not want to be challenged and generally, other than a lecture hall, do not want to take the time to explain their thinking behind their research. People with PhD degrees, especially in science, do not always possess good social skills. This is primarily because they spent their years studying and not at social gatherings. As a result, and while not 100 percent correct, they are not necessarily good communicators out of their field of expertise. This has resulted in a failure to effectively communicate with the public on matters such as the issue of global warming, the impact of COVID-19, and the number of people who have said that they do not trust a vaccine and do not plan on taking it just to name three.

Fake news

The idea of fake news has gotten traction in recent years. Claiming information is fake news questions the correctness of a statement, report, or news that someone said in the media. The word "fake" means much more than the information is not correct. It means that the information is not genuine. It may be a forgery. It may be a sham, a hoax, or counterfeit. These words have much more meaning and flare than just saying the information is not correct.

The problem here is that often the person claiming fake news has nothing to support the idea that the news is fake. The thinking is that if I continue to say the news is fake repeatedly, people will believe me without me providing support that the news is fake. This is especially important in that if the news is accurate and correct, but it has been identified as fake and that is believed the person who is claiming fake news has gotten away without being embarrassed, or worse, an offence that may even be considered criminal.

You may ask why someone doesn't challenge a person that is claiming something is a fake news. The answer is simple. The person hearing the information is fake trusts the person claiming the news is fake and does not want to take the time or the energy to check it out. It is much easier to accept the words of a person than to challenge their words. In addition, declaring something is fake news is also a way of controlling people, telling them, "Do not pay attention to the news. Listen to me. I know best." When challenged, "Why do you say the news is fake?" The answer is usually, "This is what I heard from a good source, but I do not remember their name," or "I read it somewhere, but I am not sure where nor the author." Neither are considered documented sources. The desire on the part of people making claims of fake news is to place doubt in the mind of the people. It is not difficult to create this doubt, especially in the minds of people who do not seek documentation for comments made that do not seem to be correct.

Similarly, the word "hoax" can be used instead of fake news. As a noun, "hoax" can be a malicious deception and an effort to deceive. As with fake news, people will not take the time to review the news or claim, and if the claim is said enough times, some people will believe the misinformation, instead of the truth. If the person claiming the news/information is fake or a hoax is questioned about the source that news is fake or a hoax, once again they will claim they heard the information, and they do not remember the source, but the source is well respected in their field. Of course, none of this means anything without the identification of the source and the documentation that lead to the conclusion that the news or information is fake or a hoax.

The anti-enlightenment movement

Enlightenment is an awakening. The European intellectual movement of the late seventeenth and early eighteenth centuries was an era of learning that emphasized reason and individualism rather than tradition. It was a time that the world moved forward with new ideas and great understanding of science and the arts. It was an era that catapulted life forward. It was a significant time for the world. Today there is a large group of people who identify themselves as anti-vaccine. They will not take the vaccines for any number of reasons. What these people do not know is that their objections to the vaccines are nothing new. Anti-vaccine leagues formed in England objecting to being vaccinated against smallpox in the early 1800s. The rationale for not getting vaccinated varied from sanitary reasons, religious reasons, reasons concerning science, and objections involving political reasons. Some objectors, including local clergy, believed that the vaccine was unchristian because it came from an animal. Many people objected to vaccination because they believed it violated their personal liberty, a tension that worsened as the government developed mandatory vaccine policies. This should all sound familiar. Remember that this was occurring in response to the smallpox vaccine in the 1800s. These objections surface when any new vaccine is developed. It is hard to believe what our lives would be like today if those who objected to the smallpox vaccine had prevailed. The viral disease that killed millions of people every year was declared eradicated in 1977 thanks to a massive global vaccination program.

Today we are faced with what has been identified as the anti-enlightenment movement. This is where people generate unsubstantiated, non-supported ideas that they announce are true. They seek out all information that looks like support for their thinking/statements regardless of its accuracy or peer review. As a result, they declare their thinking is accurate and correct and will not hear or accept any thinking that is contrary regardless of the accuracy or acceptance. For example: ideas such as the world is flat, Darwin's theory of evolution rather than creation and climate change, they believe are all wrong and not happening regardless of the documented evidence presented. In part, this is the fifteen minutes of fame for a member of the anti-enlightenment movement. Part of the difficulty here is that, at times, anti-enlightenment movement is being supported at the highest levels of government. This has given a renewed life to the movement. It also fosters the question, what can I believe, and why should I believe it?

You are probably saying to yourself, "I know at least one area of information that is accepted as true—mathematics. Mathematics is true, I know that 1 plus 1 equals 2." That is true, but you need to understand that the number system we and the rest of the world use, an Arabic system, is based on a system containing ten symbols that we call numbers. These symbols or numbers include 0, 1, 2, 3, 4, 5, 6, 7, 8, and 9. All the numbers you can create, write down on a piece of paper, includes one or more of these numbers in it. This is called the base 10 system because it has ten symbols or numbers, 0 through 9. Perhaps the best way to introduce a very difficult concept is to consider, when you were first introduced to numbers, you were taught only the numbers 0 through 5, instead of 0 through 9. This is a base 6 system versus the base 10 system that we were taught and presently use. Let us suppose you did not learn and have never been presented the symbols for the number 6, 7, 8, and 9. We know that in the base 10 system, 9 plus 1 equals 10. Remember in the base 6 system, you do not know the numbers 6, 7, 8, or 9. In the base 6 system, 5 plus 1 equals 10. Why, you may ask? Again, because there are no numbers 6, 7, 8, or 9 in the base 6 system.

You have never been taught them. Therefore, in the base 10 system, 9 plus 1 equals the next number you are aware of, and that is 10. In the base 6 system, 5 plus 1 equals the next number you are aware of, 10. While that may be understandable, it gets complicated when trying to do multiplication and division in the base 6 system, having learned the base 10 system. So why have I spent time on this issue? The answer lies in computers that provide answers to much of the questions of science. Computers operate on a base 2 system, 0 and 1. So 1 plus 1 equals 10 in the base 2 system. Computers, working in the base 2 system, can rapidly perform countless mathematical calculations while providing the results in the base 10 system so they may be used and understood.

Data / information as documentation

Data usually comes in a numerical form. When information is presented in a non-numerical form, it is often converted into numbers that can be statistically evaluated. Numbers are often tested using tried and true statistical techniques. These techniques allow an expert in statistical analysis to draw conclusions based on training and experience. This process is much different from a person expressing their opinions, thoughts, and ideas on a subject. Too often people believe that a nonscientist speaking on a science-based issue has the same credibility as a scientist speaking on the same issue. Nothing can be further from the truth. It would be like you going to the butcher for advice on the surgery you are planning on your shoulder because a butcher, like a surgeon, cuts meet. Truth is identified when a third party or parties, independent of a given belief or quest for truth provides documentation/information that supports a belief. The more documentation, the strength of that documentation and the creditability of expert interpretation of that documentation, the stronger the truth.

We often are dependent on expertise to interpret documentation. The law defines expert witnesses, when their testimony is important to the outcome of a trial, as a person who is permitted to testify at a trial because of special knowledge or proficiency in a particular field that is relevant to the case. However, a person's ability to testify at a trial as an expert witness is often, as it should be, challenged. It is in that challenge that the expert witness' education and experience in the field they are about to testify is confirmed, and they are permitted to provide their expertise based on education in a specific field and experience in that field.

Justice and the expert witness

The following is an example of rejecting documented information. A carpenter is building a porch for a client. The carpenter has a helper. The helper has been trained and educated in taking measurements so the carpenter can correctly cut pieces of wood so that they can be nailed to the floor of the porch. The carpenter is told by the helper to cut an eleven feet two inch piece of wood two feet, eleven and three-fourth inches to fit the opening on the porch floor. The carpenter decides not to listen to his helper and "eyes" the opening and cuts the wood. The piece of wood does not fit properly. This is an example where accurate data is provided and ignored, causing a poor result.

Conspiracy theorists

These are people that come up with ideas, theories, to blame others or to point out what is believed to be true and documented and where the conspiracy theorist say it is not true. The thinking and the theories are usually politically driven. Many times, a proven scientific truth is said to be not true by the conspiracy theorists. For example: When Neil Armstrong took man's first step on the moon, it was an unbelievable scientific achievement for him, Buzz Aldrin, who was with him, and all the astronauts, scientists, engineers, mathematicians, and others that were associated with the project. Some of the things that needed to be done included calculate the rotation of the earth and include it in the need for thrust to move the astronauts into orbit around the earth, let alone the calculation of the thrust needed to move the astronauts out of earth orbit onto a path to take them to the moon. The science and engineering to do all this was and is remarkable.

The conspiracy theorists in one sentence and without looking into what it took to land on the moon claim it did not happen. They have said and possibly believe a studio was established where the entire event was staged and filmed. There is no explanation as to why they would make such a claim, considering the overwhelming evidence to the contrary. Unfortunately, they are not interested in studying or reviewing the clear and documented evidence. They would rather just say it did not happen. Concerning the conspiracy theorist thinking that the world is flat, images taken from the astronauts clearly show the earth as a ball-like object, to which the theorist say it is all a hoax. The picture is clear evidence that the world is a sphere and not flat. The conspiracy theorist claims the picture was doctored to make it look like earth.

So why do people offer conspiracy theories? Perhaps for their few minutes of fame or to note they are "different," to be singled out. It is also believed that if you present inaccurate, incomplete, or misleading information for a long enough period, perhaps some people will begin to believe it and begin to parrot the misleading information as truth, convincing others the incorrect undocumented information is accurate and correct. If you trust the person telling you the misleading information and you do not have the time, energy, or desire to "check it out," it is easier to believe them and move on.

In some cases, people do not want to hear information, the science, that negatively impacts their life. They misunderstand that what they are hearing from a scientist is an interpretation of data from a person who is an educated expert with experience regarding the information being presented. In some cases, the science is not clear to the listener, or the listener has been inundated with information on the subject. They have heard too much information and are tuned out or turned off regarding COVID-19 information. Concerning COVID-19, one woman said to me, "Enough already. I want to get on with my life." In some cases, they blame the person supplying the information, the messenger, for the negative impact on their life. In other cases, threatening the one supplying the information with harm, or calling for their dismissal or firing. This is a sad condition in that the failure is on the part of both the presenter and the listener. While the presenter can often provide better information as to why they are suggesting disruptive action, e.g., vaccination, mask-wearing, and social distancing, the listener needs to get beyond what is happening to them and try to understand how the suggested action will benefit them and the larger audience. This is sometimes difficult since what the listener is hearing is not what is being said. They are hearing loss of job, no or little income, loss of car, home, health benefits, and more. All these are real crises and not part of a pandemic solution. They are all spin-off crises that need to be addressed and managed independent of the management of the pandemic. Perhaps the best way to address a conspiracy theorist is to say, "Thank you. Please provide your supporting documentation for your comments." Their typical approach is to raise the decibels of the conversation, explaining that the documentation that is not provided is not important. Listen to what they have to say. "Documentation is not necessary." I encourage you to thank them for their comments and explain solving the COVID-19 pandemic is critical to our global future and encourage them to join the effort as a leader, follower, or if they cannot do either, then get out of the way.

Figure 3 provides a graphic look as to what happens when science communication loses ground to misinformation and conspiracy theories. At some point, unless science somehow finds a way to communicate in a way that influences public opinion, public opinion will not believe science no matter what science communicates. This results in a serious problem.

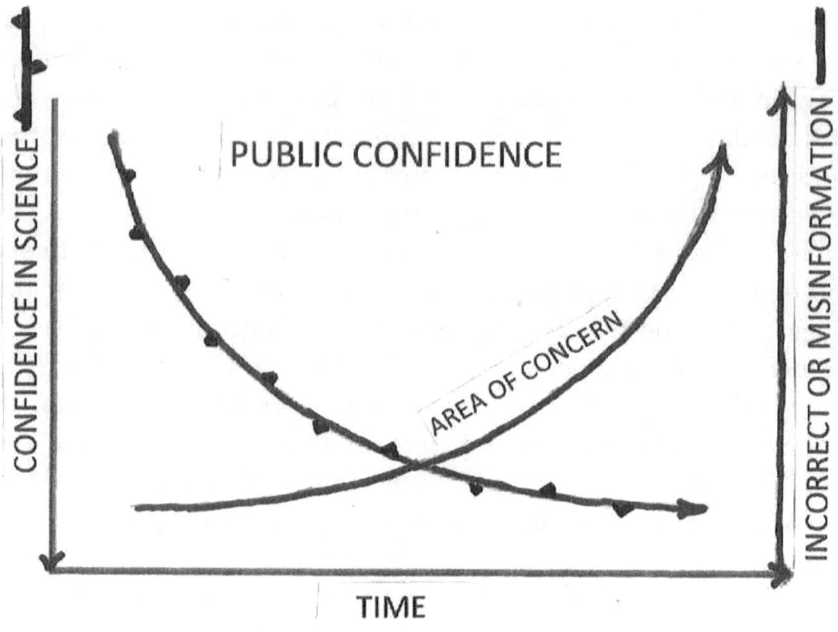

When science loses ground to misinformation

Figure 3

The bottom line with any pandemic is that there will be spin-offs or sub-crises. In some cases, the spin-off may appear to be more impacting than the pandemic. Critical to all of this is that the pandemic is paramount, not the spin-off crises. Why, you may ask? If you work on solving the crises because of the pandemic and not satisfy the pandemic, the pandemic will not disappear. However, if you satisfy the pandemic, the spin-off or sub-crises will disappear, not necessarily immediately but in a reasonable amount of time. For example: the COVID-19 pandemic has caused a severe, economic crisis. If we apply the maximum energy satisfying the economic crisis, the COVID-19 pandemic will be around possibly for years. However, if a separate group works to solve the economic crisis while the pandemic management team continues the work to satisfy the pandemic, the economy will recover, perhaps over a series of months, not years. The same thing can be said for job loss where a person cannot pay for food, mortgage, rent, their car, medicine, medical bills, to say nothing of the ramification of school closures, where one spouse must leave their job and care for the young student who will be learning from home. In most cases, the parent staying home is not a teacher and is not comfortable teaching their own children. Please do not misunderstand. All these crises must be addressed. However, the pandemic management team would best assign teams to address each of these crises and not take the focus off solving the pandemic. Remember, if the pandemic is satisfied, the spin-off crises will be satisfied faster than they will if the pandemic is not satisfied first.

The United States health agencies need to respond to conspiracy theorist in a strong manner, explaining that lack of response can be read as complacency or agreement with the misinformation and conspiracy theorists' comments. The agencies must explain that the comments of the conspiracy theorists are not supported by documentation. Only science through a rigorous process and peer review has the support that what science is presenting is accurate and correct. The words from the conspiracy theorists are only unsubstantiated worthless rhetoric.

Peer review

I have chosen to depict peer review as a microscope. It implies the level of detail required to ensure the correctness of a scientific finding.

When scientists discover something new, and he or she has completed all the documentation necessary including statistical analysis, they cannot go public and publish their findings. They must undergo peer review. This means that their finding or discovery must be reviewed by one or more competent scientists in the field agreeing that they have made their discovery to confirm that their findings are correct and deserving of publication. This is very unlike someone making a statement like, "Do not get vaccinated because you will need to have the arm receiving the shot amputated." This is stated without documentation nor peer review. Bottom line is that the statement by the conspiracy theorist, unfortunately and in all probability, will receive equal acceptability by the public as that of a peer-reviewed statement from a scientist that mask-wearing is essential considering the Delta variant. Because something is heard does not constitute acceptance for publication. The conspiracy theorist would be rejected for publication due to no documentation supporting their statements.

Comments are now surfacing that masks and vaccines should not be mandated. I would first like to address masks. The primary entrance points of the virus to the body are through the nose and the mouth. The mask blocks entry to these primary points of entrance. It is not political. It is a matter of fact and human health. In addition, people have taken the position that vaccines should not be mandated, that getting vaccinated is a personal decision. I would like to point out a few areas where requirements exist and mandates are in place. For example: a child entering public schools in the United States must be vaccinated for certain diseases like polio, mumps, and measles; or they cannot enter the school. In this country, you are required to pay taxes. That is the law. It is a mandate. To drive a vehicle in most states of this country, you must take a test and pay for a license, another mandate. For the male portion of the population, the draft is a mandate. In addition, serving on a jury is a mandate. You cannot sell controlled substances. It is a mandate, a law. Many communities have established noise ordinances which are a mandate, a requirement. You cannot drink alcohol, reach the level of intoxication, and then drive a car. It is a mandate. Many states require vehicle inspections to ensure that the vehicle is safe to drive to protect people from getting injured by vehicles that are not properly maintained. We know to stop at stop signs or red traffic lights and know not to park in areas that say No

Parking. These are all mandates that we follow with none to minimal push back. People live with mandates all the time. These mandates are designed to protect the health and well-being of the population. One must ask, "Why accept these mandates and reject a mandate to be vaccinated against COVID-19 and its various variants and to wear masks?" The vaccine and masks are designed to protect human health.

With a preponderance of evidence that vaccines work to prevent COVID-19 illnesses, 94.5 percent efficacy, with numbers of hospitalization and deaths dropping due to more people being vaccinated, with people coming down with COVID-19 again rising due to the Delta variant among those who are not yet vaccinated, with 80 to 90 percent of new cases occurring in unvaccinated people, one must wonder why people are still choosing to not be vaccinated.

Example of hospitalizations from
unvaccinated and vaccinated people

Vaccinated people who have been hospitalized with COVID-19	Unvaccinated people who have been hospitalized with COVID-19
XXXXXXXXXX	XXXXXXXXXX
XXXXX	XXXXXXXXXX
	XXXXXXXXXX
	XXXXXXXXXX
	XXXXXXXXXX
	XXXXXXXXXX
	XXXXXXXXXX
	XXXXXXXXXX
	XXXXX

Note: Each X represents one hospitalized person.

Above is a vivid example of the distribution of hospitalized people with COVID-19 that are vaccinated and unvaccinated. While it varies slightly from hospital to hospital, some 80 percent to 90 percent of people in hospitals with COVID-19 across the United States have not been vaccinated. This graphic depicts 85 percent of hospitalized COVID-19 patients are unvaccinated.

Why are people choosing not to be vaccinated, considering overwhelming evidence to the contrary? The answer apparently rests with several factors. People are choosing to believe conspiracy theories over science. People are identifying medical and science community members as lying to them. For example, "I was told that once I am vaccinated, I did not need to wear a mask, and now I am being told I need to wear a mask. Why did they lie to me?" Some people have chosen to follow a group not wanting to be vaccinated and fear rejection if they leave or do not follow that group. These people choose to not be educated in the science for fear of learning something that may change their mind. Some people believe that government should not impact their lives except for having an army, navy, and marines to protect them should the country be attacked. They look for guidance from leadership. Even if that guidance is wrong, they consider it correct because the guidance comes from leadership they support and satisfies their thinking and position regarding viruses. However, information is surfacing that unvaccinated people who are having second thoughts about getting the vaccine have used disguises that they will not be recognized, fearing that they will be rejected by the group who will not be vaccinated. The subject of religious freedom also gets involved. Some have said, "God provides me immunity. I do not need a vaccine." Without addressing this comment directly, if they believe God provides immunity, God also helped guide the hands and minds of those who came up with the vaccine that, to date, has saved countless lives. If God provides the immunity, why did *God* guide the minds and hands of those who created the vaccine if *God* did not want it to be used?

Viruses need to be considered complex chemical proteins. People typically do not see a virus being a chemical in the same way people generally have difficulty identifying food, medicines, paint, and cosmetics as chemicals, but they are all chemicals. There are examples where people hospitalized with COVID-19 have begged for the vaccine once they are in the hospital and their condition is deteriorating. The vaccine is to prevent COVID-19 illness. If you are unvaccinated and have the virus, the vaccine cannot prevent something you already have. In addition, it takes about two weeks before a vaccinated person receives the full benefit of the shot or shots, depending on the vaccine of choice. The vaccine basically blocks entrance of the virus into the cells. Masks block entrance points to the body, the nose and mouth, where virus typically enter. When people double mask, the conspiracy theorists say that is theatre. It is not. Unless you are wearing a respirator, the sides of the mask are partly open in many cases. Wearing two masks reduces the opening on the sides of the mask and provides greater protection.

Mutation of the virus is where the virus changes. It can occur because of external influences or not. This could be heat, sunlight, cold, pressure, and many more influences on the virus. We are unaware why this happens, but it happens. Sometimes the result is a more-impacting virus. The Delta variant virus is judged to be more than one thousand times more impacting than the Alpha virus, the one we first experienced in early 2020. People have said that a key scientist has lied to them when they said, "Once you are vaccinated, you do not need to wear masks," and then said, "You need to wear masks." This comment is wrong! There is no lie involved. The virus changes, and the data, please see, I said data, point to the fact that communities need to be better protected hence wear masks. The scientists would be amiss in their responsibility if they did not recommend additional measures to protect the community because of changes to the virus, e.g., going from the Alpha to Delta and now the Omicron virus.

The definition of a mutation

A mutation occurs when a DNA gene is damaged or changed in such a way as to alter the genetic message carried by the gene.

As a sidenote: in interpreting recent data, there appears to be a dramatic decrease in influenza this year. This has been attributed to the wearing of masks to prevent COVID-19. Apparently, the use of masks has also prevented the spread of influenza, the cause of the 1918 pandemic. It may come as a surprise, but the 2018 virus that caused the pandemic is still with us. The flu shot available each year is designed to prevent an influenza outbreak. In addition, each year the vaccine is modified to address the mutations of the influenza 1918 pandemic virus.

You may ask, "With all this wonderful science information and ways to document findings, why can't science do a better job of explaining COVID-19 and its variants?" This is a great question that deserves an answer. There are several answers. Sometimes, like speaking to your medical doctor, there is too much jargon. The education of a scientist occurs using jargon. The medical doctor or the scientist must communicate to the public without jargon. In addition, the scientists knows that his or her words are important and carry weight. In the case of COVID-19, the one thing the community wants to hear is all will be well and not to worry. Multiple variables in science make it very difficult to make such declarative statements. People want to learn about the virus. I have spent much time with people who do not want to be afraid, but they are afraid. Viruses do not reproduce, are not considered alive like bacteria, and there is no pill to kill a virus. You can't kill it because it is not alive. Simple explanation like, viruses enter the body primarily through the nose and mouth and somewhat the eyes. A mask that covers the nose and mouth blocks about 80 percent of the opportunity for the virus to enter the body. Some people do not cover their nose with the mask, not realizing they are not fully protected from the virus entering the body through their nose.

When the virus attacks the lungs, it attacks the cells of the lungs. The virus cannot reproduce. It cannot grow in the lungs. However, the lungs cells reproduce and carry the virus with them. This allows the virus to spread in the lungs. You may ask about quarantine. Quarantine implies that people will not be in contact with other people for a period, usually two weeks. If you assume that most viruses take two weeks to run their course, if someone with COVID-19 virus is quarantined for the time period that a typical virus takes for a person to recover, the person who quarantines should not cause the virus to spread to other people.

CHAPTER 7

The Difficulty and Importance of Testing Science to Ensure the Communication of Accurate Information

It is difficult to understand what goes into a science effort to develop something that no one has ever discovered. I want to spend time on the development of experiments and studies necessary to determine the accuracy of experimental findings. It is important to understand the rigor necessary to ensure the proper study is being carried out and the information/data and its communication. This is being done to provide some insight into how information/data is created and handled such that it may be evaluated and statistically tested so that the findings may be communicated with certainty and accuracy. This is what needs to be done in developing a PhD thesis and what is needed to be done in creating vaccines for viruses, including COVID-19. Compare this approach with the people claiming that science cannot be trusted and their comments are not documented nor substantiated.

Experimental design:

- What is being studied, and why?
- What is needed to carry out the study?
- When can the work begin?
- Where (the location) can the work begin?

- Why is the study being done?
- Who will interpret the data? Who will peer review the findings? And who will communicate the results?
- How will the interpretation of the finding be shared, and by whom, e.g., technical paper, seminar, other?

Since I was not involved in the creation of a vaccine, I cannot speak to the design, planning, testing, and the like for the vaccine. However, I can identify what I needed to create as an experimental design to satisfy my thesis for something that had not previously been done. Experimental design is often overlooked, while it is critical for obtaining key information needed to acquire meaningful data/information that can be evaluated/interpreted and communicated. Before I decided to conduct an experiment, such as the reason masks work, I needed to understand what I wanted to study and why I wanted to study it and evaluate what I was trying to prove or disprove.

For example: when I was trying to determine a topic, I had to find something that no one had ever done before. I read an article from a scientist that said vegetation is not impacted by air pollution unless and until the vegetation shows a physical impact, like browning of their leaves. This seemed strange to me in that there must be something going on in the vegetation biochemically since the brownness of the leaves did not just happen. In addition, this change in the leaves must impact other things such as growth before the leaves would brown. I decided to focus on the impact of air pollution on vegetation. However, this is only the beginning. Next was how to create an experimental design to allow me to present the idea that vegetation is impacted by air pollution before they demonstrate a browning of leaves.

With the identification of what is to be studied, I needed to design an experiment that would provide data to demonstrate and satisfy my thinking/my thesis. Therefore, I needed to

- select the type of vegetation to be used
- select the air pollutant to be used in exposing the vegetation
- a means of measuring plant growth

- a plan that would permit exposing the vegetation to different concentrations of an air pollutant and measuring the response
- determine what if any nutrients will be used

I chose oat seedlings because the plant grew straight upward and could be easily measured. The selected pollutant was sulfur dioxide, a pollutant that was very prevalent with the burning of coal and oil. I used a simple measuring devise to measure the height of the plant. I first though of a simple ruler that would measure inches. While this would work, the tick marks on the ruler were too large. I could measure 1/8-inch intervals. So I had 1/8, 1/4, 3/8, 1/2, 5/8, 3/4, 7/8, and 1 inch as a measurement. I decided this devise was not able to measure the small change that I anticipated. It had eight units of measurement. I chose a metric measurement where I had 25.4 measurement units, millimeters (mm) per inch. The use of a metric measurement permitted a more-exacting measurement. The plan was to place a vegetation in a plexiglass chamber, where I could introduce sulfur dioxide into the chamber through a metered device. Since I calculated the volume of the chamber, I would release sulfur dioxide into a chamber until the chosen concentration level was reached. The plan was to begin with low concentration levels of sulfur dioxide and increase the concentration and measure the height of the vegetation after the passing of several hours. After twenty-four hours, I would stop the experiment, clear the chamber, and select a new sulfur dioxide concentration, new oat seedlings, and expose the vegetation to the new concentration.

A series of experiments were done without sulfur dioxide to get growth norms in the chamber. Growth data was collected over a series of weeks, exposing the vegetation to different levels of sulfur dioxide. The experiments stopped when a concentration level resulted in the vegetation leaf turned brown. Therefore, if there were changes in growth with sulfur dioxide present prior to the leaf turning brown, it was affected by sulfur dioxide. And if there were no visible changes to the vegetation, the leaf was not brown.

While all looked well at this point, I collected my data and my interpretation that growth was affected at concentration levels below the level where the leaf turned brown. Further, I was able to determine that at the lower concentration, the growth rate slowed with the introduction of low levels of sulfur dioxide, and after a period of time, vegetation growth rate began to slowly increase.

With all this, I felt good that I had proven that the vegetation I used in the experiments were negatively impacted without any outward signs of distress such as the browning of leaves. I decided to meet with colleagues to discuss my findings. When we met, one of the members of the group said something I shall never forget. He said, "I like you, but I do not believe you." I was puzzled by his comment and did not know exactly what he was trying to tell me. Apparently, he saw the look on my face and suggested that I come to his office. When I arrived, he explained to me that I needed to employ statistical analysis to prove what I was saying. He suggested the t-test. His meaningful comment was when he said that if I could support my findings with the t-test statistical analysis, and he said he thought I could, he and others could no longer question or challenge me and the claims I was making concerning my findings. They needed to question and challenge the results of the t-test, and they could not do that. I did exactly what the statistician suggested, and the results identified an impact of sulfur dioxide on the growing vegetation. As I increased the concentration of the sulfur dioxide, the longer it took for the vegetation growth rate to respond and return to an expected growth rate. When the concentration of sulfur dioxide reached a level where growth did not take place, the leaf slowly began to turn brown. I point this out as an example of what is needed in terms of scientific rigor before a scientist can make a claim. It must be fully supported before it can be communicated. This is unlike those who choose to "shoot from the hip" with unsupported comments about COVID-19 and its variants.

The fundamentals of data management: the difference between information and data

The question is sometimes asked, "Is this data, or is this information that we are calling data?" Data are numerical while information is generally not numerical. For example: If you wish a list of students presently attending the fifth grade at your local grammar school, acquiring the list can be identified as information. If you further asked for their average age of the named students, their grade point average, and the results of their IQ test, the results are considered data.

Data identification

You must identify the type of data you need so that the evaluation you seek can be carried out. If you want to determine if the students presently attending the seventh grade have better grade point average than the seventh-grade students over the last three years, you need the test results for all students in the seventh grade for the last three years. While this review of data examples appears simplistic, and it is, it clearly identifies the objective and the data needed to carry out the review.

Data collection

With the objective for the data to be collected understood and the type of data to be collected identified, there is a need to identify the sources for that data. There may be more than one source to be collected. For example: if the objective is to determine if a community needs to build a football stadium that will also be used for baseball, that would require the identification of one or more sources of information. If it is possible that two stadiums should be build one for football and one for baseball, additional data will be needed.

Data sources may be acquired in several formats:

o personal interviews
o existing files

o   physical inspection of facilities
o   other sources

The data may be received via:

o   electronic transfer
o   hard copy
o   notes from interviews and physical inspection
o   voice mail recordings
o   others

Data interpretation

This is the most difficult, most important, and most valuable part of the process. Keep in mind data are typically a series of numbers that must be evaluated by knowledgeable individuals/experts who will determine if conclusions can be drawn from what has been collected and tested. The evaluator is looking for possible trends, unexpected spikes, or the lack of expected trends or spikes.

Often statistical analysis may be employed to help the expert support conclusions or the lack of conclusions. This may include:

* T-test analysis
* Analysis of variance
* Chi-square analysis
* Standard deviation
* Correlation coefficient
* F-test analysis
* Among other statistical approaches

The twelve-month moving average

One example of a simple technique that may be used in analyzing data is called the twelve-month moving average. This analysis removes year-to-year variables that may impact conclusions. The technique involves collecting numerical data for twelve continuous months, January through December, adding the numbers together and dividing by twelve. This provides what is referred to as an annual average. The twelve-month moving average technique requires the number from January to be removed and replaced with the numerical data from January of the following year. Therefore, the numerical values from February 1 of the first year to be added through January 31 of the second year; and the sum of those number are divided by twelve to acquire an annual average, but the year is from February of year 1 through January of year 2.

This approach of dropping the numerical value of the previous year and adding the numerical value of the following year creates a twelve-month moving average. This becomes extremely important in determining trends because the approach has removed possible variables when trying to compare year-to-year results. For example: If a company's annual Occupational Safety and Health Administration's (OSHA) total recordable injuries suddenly increases from year 1 compared to year 2, a closer look uncovers that ice and snow not seen in years resulted in four automobile accidents in year 2. This compares to zero automobile accidents in year 1. While the numbers cannot be denied, they have resulted in a doubling of the company's total recordable rate from year 1 to year 2 that has significantly negatively impacted the company's progress in improving safety performance. If the twelve-month moving average is employed, the impact on the company safety performance will be buffered because the unexpected ice and snow did not represent the norm. While the data cannot be eliminated, it will be used in a manner that distributes the data across a twelve-month period and ensures that the ice and snow does not significantly impact the company safety performance. This is an effective way to use a technique that does not eliminate the snow and ice but spreads it in a way that it may be viewed over a longer time period. By using a twelve-month moving average, ice and snow from year 2 is less of a factor.

Data communication

Meaningful communication of data providing the results of an evaluation is critical. Too often those providing the conclusions of the data interpretation provide information that is far too technical for the conclusions to be understood. The report should include:

1. an executive summary
2. an introduction
3. data collection
4. data evaluation
5. conclusion(s)

Simple communication with the least amount of jargon works best. Too often scientists are unwilling to make simple conclusive statements for fear of being challenged regarding the simplicity and the lack of details. In addition, the details that lead to the interpretation and the conclusions are based on statistical analysis that are often complex. For example: a picture of a cigarette an equal sign followed by hundreds of cans of diet soda send a clear message of the impact of one cigarette without saying a word. The lack of formulas, graphs and charts would strike fear in the hearts of some scientists, but for the consumer without a science background, the information received is clear, to the point, and appreciated.

T-Test X analysis: rejection of technical jargon

In communicating the results of a data interpretation, evaluation, and conclusions, one must first know the audience who will receive the information gleaned from the data. Know the expectations of that audience. In some cases, age, ethnicity, gender, and other influences will result in affecting audiences differently.

The fundamentals of data management

You have received the data you require for analysis, evaluation, interpretation, and conclusions. Before you do any analysis, the data needs to be organized/managed so that the results and conclusions are meaningful. Again, you must consider your audience for the conclusions you will be drawing. For example, if you are reviewing test results of students in your community, you may want to consider grouping the students in clusters:

- Male students
- Female students
- Ethnic backgrounds
- Ages
- Others

Who is the data for, and why?

Here we are back to the audience for data. In some cases, the audience is paying for the results and subsequent conclusions. When a third party is paying for the analysis/evaluation, the third party should not be permitted to skew the collected data to influence the results and the conclusions. If there is no audience for the results and conclusion, the question should be asked, "Why is the data being collected and the results evaluated and conclusions drawn?" If annual studies are performed and the report of the findings and conclusions sit on a shelf, the question of why go through the expense of collection, analysis, conclusion, and a report when no one reads the report.

The who, what, why, where, when, and how of data collection:

Words such as who, what, why, where, when, and how are often used to get a better understanding of a topic.

For example:

Who needs data? Any person or group wanting to learn more about an ongoing situation or even when the information from that event will ensure a better understanding and, as a result, hopefully permit growth and use of conclusions will result in enhancement or improvement.

What is the reason for data collections, analysis/evaluation, and conclusions? It is to learn, improve understanding, and as a result, employ the conclusions to create a path forward.

In some cases, data is gathered to identify and determine a level of risk. Entire texts have been written about risk.

Incentives for data-gathering. The primary reason/incentive for data gathering is to improve understanding and hopefully arrive at conclusions concerning an issue, an event, or a risk.

So, what are the fundamentals of data management? It is the collection of accurate and meaningful data to use that data in an experimental approach designed to enable an evaluation to bring about meaningful conclusions.

Providing Health, Safety, and Environmental (HSE) data to government leaders and decision makers: First and foremost, you need to realize that senior government decision makers are facing multiple complex issues that need decisions. While they need to know the conclusions and possible recommendations of the experimental evaluation, their time is short. Please keep in mind that there are others vying for their time, also with conclusions from work that they have undertaken. You need to be short and to the point. If it takes more than ten minutes for the complete presentation, it is too long. You need to prepare for a presentation of a maximum of ten minutes. If senior staff wishes to learn more, they will ask questions.

Further, it is best to prepare a one-page briefing paper for senior management to review prior to a presentation. The briefing paper should have four major headings. These include:

- What is the issue?
- Why is it important?
- Where do we stand on this issue now?
- What is recommended?

With a briefing paper and the less-than-ten-minute presentation, senior management should have all that they need to understand the issue, ask questions, and/or come to a conclusion.

Types of data evaluation:

- Evaluations designed to provide information.
- Evaluation designed to provide information calling for change.
- Evaluation designed to provide information calling for no change.

    Data collection and grouping. Often as data are gathered, they need to be grouped to avoid confusion and misinformation.

This is especially important in gathering data during time of a pandemic or crisis. When a pandemic or spin-off crises occur because of a pandemic, and is clearly identified, those responding must stay focused. It is not impossible to see where new data surfaces during a pandemic or crises that has the possibility of taking the focus off the pandemic or resulting crises and on to what is perceived as a "second crisis." This new data that have the possibility of refocusing the pandemic or crises needs to be separated out so that the focus on the pandemic is not lost. Example: There is an explosion and fire. People are missing and thought dead. Firemen are fighting a huge chemical fire. A rescue vehicle coming to the crisis went off the road and crashed into a grammar school. Children are feared dead. It is easy to see that focus on the chemical fire can shift to the accident at the grammar school. The focus must not shift. Resources will be provided the school, but the accident must not cause the crisis focus to shift. In the case of a pandemic, business closures resulting in job loss and loss of income must not take the focus away from addressing the pandemic. Job loss and loss of income are very important and must be addressed by crises spin-off sub-teams and must not be the focus of those addressing the pandemic. In some cases, political issues get in the way of what needs to be done to address a pandemic. This is the worst of all worlds. This is when politics interfere with what needs to be done, resulting in additional deaths and a prolonging of a pandemic. When this happens too often, science is put into a defensive posture and shifts its role to one of defense. This results in an unnecessary waste of time, money, and most importantly, lives. Politics have no place in a pandemic.

The selling of data

The selling of data does not mean to receive money for information/data. Instead, it means the ability to convince people that the data contained in the evaluation may be interpreted as a call to action. That call to action can be a news report to the public about a pandemic and subsequent crises. It could mean a need to provide masks, ventilators, or the like. It could be a hurricane approaching where the expected wind speed, storm surge, or time of storm arrival are key to saving lives and property and must be communicated to the public in a rapid and meaningful manor.

Challenging data

The use of statistical analysis in reviewing data: For most people, unless you have been trained in statistical analysis, this is very confusing. To begin, let us look at the fundamentals of statistical analysis. A better way to say it is "statistical analysis for the non-statistician." Perhaps this will help. When I was beginning my work to evaluate data, I was developing information that was pointing me to a conclusion. I previously stated that a statistician put me on the right track. To cut through this, what he was saying is that the mathematics in the t-test identifies "normal" expected results with a 95 percent to a 99 percent level of confidence. If I could use my data and the t-test to show that my results were outside the expected norms, he would have to accept them since I was no longer drawing the conclusion. The t-test was drawing the conclusion. Fortunately, I was able to do that, and my thesis was accepted.

Therefore, statistical analysis can permit an evaluation of data and compare it with expected norms. It also can predict outcomes.

For example: if I had ten pennies, and I tossed them in the air, when they all came down and settled, there would be a combination of head and tails. The probability of seeing some heads and some tails is much greater than seeing all heads or all tails. On a more-sophisticated mathematical level, mathematical models can be created. These models are designed to predict the future. For example: if I took one gallon of 100 percent pure orange juice and poured it out into a moving river, the model with the proper input could tell me how many miles downriver I would have 3 percent orange juice. While the example is not a good one, if a toxic gas was released from a hundred-foot stack at a chemical plant, the model could tell me with the proper input involving wind speed and direction, among other inputs, where the chemical would reach the ground and the concentration of that toxic chemical when it reached the ground. As a result, I would know what areas to evacuate and how much time I had to do it. While models are not always extremely accurate regarding a release of a toxic gas, I would use the results and take all appropriate action.

Pandemic and crisis communication

This is critical. No one wants to hear bad news, and no one wants their life impacted by limiting movement and freedoms. However, many times it is required. The question is how to present information of change in a manner that while not liked is understood and will be adhered to. This can only occur when the information is presented in a meaningful manner. Leadership, such as government, will be less effective by just mandating, with fines and penalties, although this may be the only way to be successful. Case in point, with the COVID-19 pandemic raging, there is a call in many parts of the country to wear face coverings, masks. Many are refusing to wear masks, saying that it is an infringing on their personal freedoms. In addition, they point to the fact that scientist in the government were saying that once vaccinated, there is no need to wear masks. Later those same scientists said there is a need to return to wearing masks. As a result, some said that "you cannot trust the scientists. They do not know what they are doing. They say one thing and then change their mind." This is a serious problem. Science needs to take a step back and identify a way to better explain the need for wearing a facial mask. In this case, the Delta variant and the fact that it is more than one thousand times more impacting than the original Alpha variant played a very large part in the government's change of position concerning the wearing of masks.

We are now faced with the Omicron virus variant. This is another variant of the COVID-19 virus. It was first identified in South Africa and rapidly made its way across Europe and the United States. People who have been fully vaccinated, and even those who have received the booster have been affected by the Omicron variant. Those who have not been vaccinated need to get vaccinated as soon as possible, considering the aggressiveness and potential impact of this variant. While the Delta variant was and is aggressive, the Omicron variant appears more aggressive than the Delta variant. Within a period of two weeks, the Omicron variant has gone from a variant in need of watching to the dominant virus variant in the United States. Those who have been vaccinated and impacted by the Omicron variant have fared far better than those unvaccinated. Hospitals are becoming overrun again, and the gathering of people during the 2021 holiday season is expected to result in another spike in cases.

The virus enters the body three ways: one via the nose; two, through the mouth; and three, and believed to a lesser extent, the eyes. Simply stated, if the virus, which kills, needs to be stopped from entering the body, there needs to be a barrier to prevent the virus from entering the body. A mask is a barrier and blocks the virus from entering the body through the nose and the mouth. A face shield or goggles, or even a person's glasses, will help reduce the virus entering the body through the eyes. This should be so obvious no one should go without a face covering.

Let us try it a different way. It is 8:15 p.m. on a hot summer evening with the temperature at eighty degrees Fahrenheit, and the low for the evening is expected to be seventy-eight degrees Fahrenheit. Suddenly you hear a loud crash, and you look up, and the picture window in the living room of your home has been shattered by a baseball. There is glass everywhere. There is now a large hole where there was once a window. The hole is large enough for a person to walk through by, stepping through the garden area. You are concerned about a possible break-in while you, your wife, daughter, and son are sleeping. What do you do? Immediately you think about blocking entry to you home with wood or something to keep your family safe. You want to block entry to your home to protect your family, not unlike getting masks for you and your family to cover the entry points of the COVID 19 virus. No difference. Both are designed to protect you and your family.

Peer review

The term "peer review" is not well understood. We come back to what is true and correct. If I develop new thinking that I want accepted, I announce my comments. I say it is correct and publish my thinking in a newspaper or some other form of nonscientific written communication. In addition, I take a force-full approach in announcing my comments. The idea being my force-full approach will result in convincing my audience that I am correct. I have offered no proof. I only raised my voice in stating my position.

When this happens in the science community, there is a procedure that must be followed. I would need to explain my position and submit it to a science journal for publication. The journal would send my submission to a person who would read my idea and examine it for correctness, documentation, and support. If, after peer review, my submission is found not to be correct, my submission will not be published regardless of how loudly I speak. This is important because a science idea must undergo peer review before it is announced and accepted for publication and dissemination. This is unlike other claims that are not science based.

The importance of proper testing, e.g., randomized placebo-controlled trials versus nonrandomized placebo-controlled trials, "the gold standard."

You will hear, "This drug has been tested and may be used by physicians to combat a disease." Suddenly there is a challenge to the quality of the test. People will claim that the results were "perfect," and the drug should be used. Obviously, there are tests concerning the quality of the drug. This results in confusion, frustration, concern, and often anger. For example: If the drug is a cancer-treatment drug and you have a child suffering from cancer, this drug will help in a possible cure, according to a test. You will do everything in your power to get the drug for your child, quoting the study that presents findings that identifies the drug as effective. This is a difficult time to say the least. The physician tries to explain that the study that you are identifying did not use the gold standard test and may cause more harm than good.

The MD degree and the PhD degree

There are doctors, and there are doctors. For most people, the word "doctor" means a medical doctor or physician. There are many types of physicians as there are many types of doctors. A person who has received a doctorate degree has the distinction of having completed the highest level of education in their chosen profession, whether that be the many different types of law, medicine, the humanities, religion, and science, to name a few.

During a pandemic, such as the one experienced and 1918 and 2019, 2020, and 2021 to date, the medical community—such as nurses, technicians, and medical doctors—care for those impacted by the virus. Others who have doctor degrees, PhD degrees in epidemiology, industrial hygiene, virology, biology, chemistry, and physics among others have the function of identifying ways of avoiding exposure to the virus.

Communicating the science of biology/medicine

The study of biology is broad, including many disciplines such as anatomy, phytology, microbiology, toxicology, among many others. Medicine combines biology, chemistry, biochemistry, toxicology, among others. During the 2019, 2020, 2021, and possible 2022, the COVID-19 pandemic and subsequent crises caused and continue to cause communities to be frightened, upset, and concerned for themselves, their families, and their children. They need concrete documentable advice. They do not need it sugarcoated or watered-down information. They do not need a cheerleader telling them all is fine when it is not. They need the truth. They need to be "leveled with," and they will respond. In other words, "Give me the facts, and I can deal with it."

Science sometimes tells us what we do not want to hear, but that is okay.

Any good manager will tell you, "When I know the facts, I am able to deal with it, plot a path forward, and work to measure progress to satisfy the issue no matter what it is and what I need to do." It is misinformation that creates starts and stops in working for a solution. For example: regarding the coronavirus, COVID-19, pandemic of 2020, there are several things known to work and other things that are known not to work. From studying previous pandemics, we can easily identify that social distancing, washing of hands, wearing of masks, wearing of gloves, among other things, can and will have a positive impact in lowering the number of cases of the disease. I want to stop for a minute and look at what I just said:

- We, as a world, have experienced previous pandemics not unlike the coronavirus, COVID-19.
- First question, "Can we learn from what was done during previous pandemics to help us with this pandemic?" The answer is a resounding YES!
- What did we learn? We learned that social distancing, washing of hands, wearing masks, and gloves work to lower the number of pandemic cases.
- So why isn't everyone following what is known to work to overcome the pandemic? I view it much like pounding a nail into the wall with the palm of your hand instead of using a hammer. Some people will continue to use the palm of their hand regardless of the evidence that it is dangerous and wrong; and more importantly, the hammer is quicker, safer, and saves time.

- People do not want to be seen as wrong, so they will not stop pounding the nail with the palm of their hand, cutting their hand as they continue to pound the nail. In addition, perhaps they did not take the time to understand the reason to use a hammer.

- Perhaps science has not reached these people with an overwhelming amount of evidence as to why using a mask, social distancing, among other preventative measures, is better than getting the virus.

- It is critically important to make the point someone supporting the use of masks, social distancing, and the like is not coming from the same position as the person that says, "Masks are unnecessary and infringe on my rights as a citizen of the United States of America." One is giving an opinion based on no supporting evidence, and the other is reporting on science-based studies with an overwhelming amount of evidence.

- Therefore, it is not a one-on-one debate. It is one-person, challenging science with no evidence to support their position versus science, a stand-alone documented and supported position. Science wins!

- Others have taken the recalcitrant position of "Do not confuse the issue with facts. I have made up my mind and will never be forced to take the vaccine."

CHAPTER 8

Behaviors Prior to, During, and After a Pandemic

Behaviors prior to a pandemic

Behaviors that may lead to pandemic fall into many categories, which include individual human behaviors, multiple interactive human behaviors, and cultural behaviors, among others. Regarding individual human behaviors, the topics of attitudes, beliefs, and feelings come to the surface. All three impact behaviors, and all three are interconnected, but not all interdependent. Webster defines attitude as "a way of looking at things, or a manner." Attitudes are generally present in a person by the age of five to seven years old. These attitudes are based on experiences and learnings from family, friends, or playmates at school. Prejudices are learned and create attitudes at a very early age. It is from attitudes that food preferences are decided and likes and dislikes are established. Attitudes are at the core of a person's existence; they are difficult to recognize, and they are very difficult to change. Webster defines belief as a "conviction." Beliefs are also at or near the core of a person and may change more easily than attitudes. The expression "I feel that is the case" and the expression "I think that is the case" should not be viewed as the same. "I feel" and "I think" are typically incorrectly used interchangeably. "I feel" is a much stronger wording than "I think." "I feel" is a conviction. Feelings and thoughts are not the same.

Webster defines emotions as "feelings." Feelings are typically one word, like happy, content, angry, and frustrated. These are all "feeling words." If you can substitute "I think" for "I feel" and it makes sense, you expressed a thought or an idea and not a feeling. For example: if you said, "I feel that we should do that" and you can substitute "I think we should do that," you have expressed a thought and not a feeling. You do not say, "I think happy." You would say, "I feel happy," so happy is a feeling. The attitudes, beliefs, and feelings of an individual influence their behaviors, and these behaviors may lead to a sub-crisis situation where a pandemic exists. A lack of focus due to distraction or repetition may result in a "human behavior" that may lead to a sub-crisis of a pandemic; for example: a lack of focus during a pandemic due to fear could easily lead to a loss in judgement that may lead to sub-crisis such as a fire, flood, or worse.

Failure to follow directions is another human behavior that may lead to a sub-crisis; however, this is often overcome using checklists. Pilots, for example, use checklists prior to takeoffs and landings. Other examples of human behavior that may lead to a pandemic sub-crisis include anger, frustration, and daydreaming. For the most part, daydreaming usually occurs when people are doing routine function and focus on something other than what they are doing. For example: workers prior to a holiday or vacation or within the first few hours after returning from a holiday or vacation often lose focus because of "daydreaming." It is during these times that the worker typically does not focus on what they are doing; instead, they focus on what they plan on doing or did during holiday or vacation rather than focus on the job at hand. This behavior often leads to injury. Other types of behavior that may lead to a sub-crisis include procrastination, arriving at work with your mind elsewhere, lack of training, conflicting priorities, stress from a pandemic, a clown behavior, and many others.

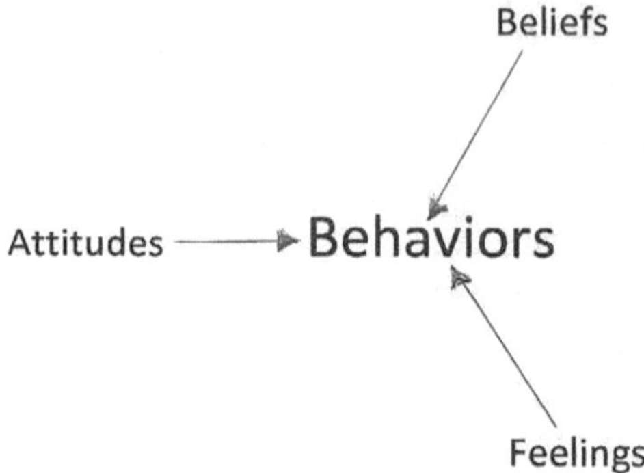

The importance of behaviors during a pandemic

Multiple interactive human behaviors that may lead to a sub-crisis situation include horseplay in the workplace, an employee's anger due to a disagreement with management, employees who are angry at each other, group labor action without thinking of possible sub-crisis consequences, and distractions such as watching the World Series instead of instrument panels in a control room, or multiples of the above behaviors and others.

Many systems are built today with redundant safeguards to prevent failures causing a pandemic or sub-crisis. As a result, failures should not occur; however, the airplane industry has experienced failures even with redundant flight systems, but these systems have proven to be very effective with incidents and accidents kept to a minimum.

With all the above in mind, how can a pandemic or possible sub-crisis be prevented? One way is through training. This includes training in the classroom, hands-on training, and discipline as key factors in preventing a pandemic and sub-crisis that may result from the pandemic. This is about the mental discipline of doing it right the first time and knowing why a person is doing it correctly the first time. It is knowing the consequences of actions before they are done, or not done. This is critical in preventing a pandemic or a sub-crisis. If someone knows the consequences of incorrect actions and behaviors, they will be more focused on performing the correct actions.

Knowledge is important to preventing a pandemic and sub-crisis. For example: where there are risks and if someone understands the ramifications of those risks, they will work to prevent an event that may lead to a pandemic or sub-crisis. It is also important for a person to realize that they should get expert guidance when and where needed. People should be able to say, "I don't know, and I need help," rather than acting where there are unknown consequences. The following are examples where proper guidance and training did not take place and a crisis resulted.

A large block of sodium submerged in oil entered a manufacturing plant and was transferred from a rail car to a vessel where it was also submerged in oil. The oil was heated by steam coils, and over time, the steam line, which was not stainless, rusted and developed a leak. The steam quickly turned to water, came in contact with the sodium, and hydrogen gas was released. The hydrogen found an ignition source, and an explosion and fire resulted, which destroyed the plant and building. A second example of improper guidance and training concerned a product transfer that resulted in a crisis. A new employee was asked to open a valve and send a batch of finished product by gravity from one tank to another. A second employee went to the floor below to monitor the flowing liquid. This second employee did not see any indication of flow on a meter and believed the flow meter was malfunctioning. The employee on the first floor tried to make the flow meter work by tapping on its side, and when this did not work, he returned to the upper floor where he learned that the first employee had turned the wrong valve and sent three days of production to a sewer. The first employee was fired; however, in truth, as with most accidents, management needs to look at itself and their practices and actions. In this case, it can be said that the employee did not receive adequate training concerning which valve to open, and management should have been held responsible. Also, the new employee was not made aware of the consequences of turning the wrong valve, for both the company and the employee.

Like the examples above, when faced with a potential pandemic information concerning quarantine, social distancing, and mask-wearing, it is critical to understand the value of the information. Information from experts will help in offsetting possible sub-crises such as shutdowns resulting in job loss, loss of income, and foreclosures because of a pandemic

Behaviors during a pandemic and subsequent crises

Behaviors are critical to managing a pandemic and their spin-off crises. In many cases, the outcome of a pandemic can be impacted or spin-off crises delayed from reaching stability due to behaviors. Emotions during a pandemic and subsequent sub-crisis, such as fear, concern, and frustration, among others, can be a roller coaster causing changes in behavior. Typically, emotions surrounding pandemic begin with shock. Questions are asked, such as, "What happened? Why did it happen? Where did it happen? Who is involved, and when did it happen?" "He was fine two days ago. He said he had a little cold." Initially, there may be disbelief. "This could not have happened. The information must be wrong. I do not believe what I am hearing. We have systems in place to prevent this from happening." "I know the people, and they are well trained to prevent such a happening." Reality soon takes over, and the information brings a sense of truth. Learning of someone who is in the hospital near death from COVID-19, or a family member who has just been diagnosed with COVID-19 brings a reality front and center. Acceptance soon follows. There is a knot in your stomach, a deep sense of loss, a fear of what is next, a sense of helplessness, and a feeling of hurt.

Emotions are high at the hospital where there is a steady stream of patients with COVID-19 while those providing care to the afflicted are becoming stressed. The refrigerated trucks housing bodies that succumb to the COVID-19 virus because the hospital morgue is full is a grim reminder of what is happening. The failure of family members to see, talk to, or touch their loved one in their final hours or minutes is adding to an already difficult time. Respect must be accorded to both the dying and their family members during this most difficult time.

However, information needs to flow. Learnings must be shared so that effective treatments and drugs may be passed not only across hospitals but with health centers across cities, counties, provinces, states, and countries around the world so that the pandemic and spinoff sub-crises may be effectively and efficiently managed. The protocol, as with any pandemic or crisis, must be, save lives first! It is not unusual for complications to occur. For example: if relatives or friends of those in hospital want more than touching their loved one's fingers through glass, the management at the hospital are in a difficult situation. They must be sympathetic and empathetic to all involved. The hospital gathering or transfer of information will become more difficult. The hospital management needs to have a plan for communication with all involved and seek information on loved ones. It is better if that plan includes meetings with all interested parties at a site other than the hospital. It can be at a nearby hotel. However, all appropriate measure must be taken to protect the hotel staff and all those coming to hear from hospital management. The plan should be to meet with all families as a group and then privately. The hospital should share information from those in the hospital to relatives trying to learn what is happening. Hospital staff should take all letters, notes, and the like from the families that the families want read to their loved ones in the hospital. Granted, this will take a major effort on the part of the hospital staff. However, it will be appreciated by those seeking information and wanting to inform those in the hospital that they are there for them.

There is such a thing as emotional intelligence, and its definition is the ability to behave in a manner that permits the taking of informed decisions that allows for the inclusion of people's beliefs, attitudes, and feelings. Emotional intelligence becomes a factor during a pandemic situation. Many times, unpopular decisions are necessary and appropriate. For example, using of a refrigerated truck to hold bodies of those who died from COVID-19 because the hospital morgue is full.

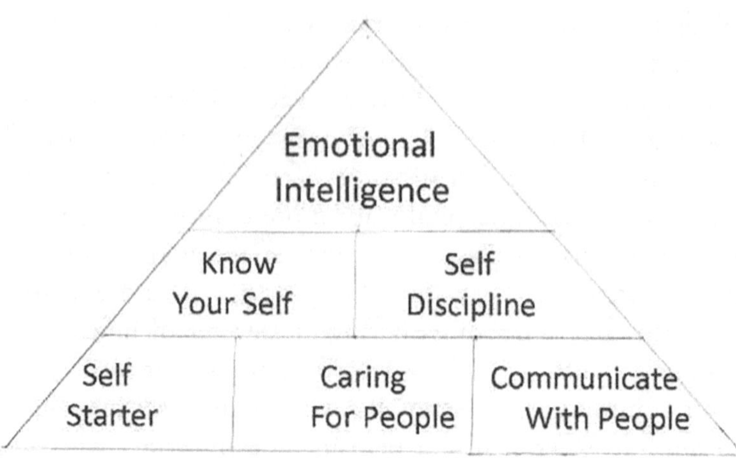

Addressing emotional intelligence

The pandemic management team members need to be trained on how to move from emotion to acceptance so that they may get on with the business of managing the pandemic and subsequent spin-off sub-crises; however, this is not easy. They must focus on the task of implementing the pandemic management plan. Some of the emotions that may be experienced by members of the pandemic management team include feelings of somber, fear, and concern. They may also experience anxiousness, sympathy, and empathy, among other feelings or emotions. While difficult, the pandemic management team must try to move beyond feelings to be effective. Emotion must not drive decision-making. Expectations of the pandemic management team include addressing of the pandemic and spin-off sub-crises that occur, fixing the pandemic and sub-crises, being well trained, giving confidence to those experiencing the pandemic and their loved ones, and working together as a team to bring the pandemic and/or sub-crises to stability. In addition, the team is also expected to work with other government regulatory agencies to solve problems.

The pandemic crisis management team will be faced with distractions, such as complaints, demands, threats, inputs from sub-crisis situations, which will take place ancillary to the crisis, accusations, and unexpected changes in the pandemic or sub-crises, among others. The team must also deal with judgments, and for the most part, they will be hard and difficult judgments. If the team is not able to function in a straightforward, dedicated manner, they will not be effective and in some cases will be too slow in their response, not focused, or not able to carry out the pandemic or sub-crisis management plan. In addition, they may not be able to take decisions and satisfy the needs of people. They must be nonconfrontational, and they must be productive. The function of the team includes teaching goals and objectives and target setting, guidelines, skill-set development, procedures, encouragement, policy, standards, practices, and programs development, and more managing the pandemic.

Activities of a management team

Most pandemic management team members have many demands and responsibilities, and therefore, they may not always be available at times of a pandemic. For example: what happens if a pandemic team member, or their designee, cannot make it to a pandemic management team gathering, or they are not available to call in and be part of a pandemic team meeting? In some cases, the pandemic management team member may not have read the pandemic or crises management plans, forgotten the training, and they do not know what to do. They may ask themselves, "What happens if I make a mistake or if I am the only member of the pandemic management team who is vocal? If I disagree with the team's selected path forward, if I disagree with the messages being prepared for the media? What will be the consequences if the pandemic management team fails in its effort, and how will my colleagues view me if we fail or if I fail?"

The pandemic management team must undergo pandemic management plan training and behavioral training as part of their preparation to respond to a pandemic. The team needs to understand the impact of attitudes, beliefs, and feelings of its members during a pandemic situation and how they may behave during a pandemic or addressing spin-off sub-crisis situations. Behaviors will vary and change as one's attitudes, beliefs, and feelings are impacted. For example:

Your parent or your child has COVID-19 and is near death. You want to run into the hospital to see you child or parent. You are not thinking about vaccines or masks. You want to tell your loved one that you love them. You want to hold their hand and embrace them. You want them to know that they are not alone. The hospital, on the other hand, forbids you to do this. While this is changing and is expected to continue to change over time, nevertheless, the issue is highly emotional.

At time like this, it helps to roll back to basics. If a pandemic and/or spin-off sub-crises policy exists that reads, "Save lives first, help families, and ensure government continuity" in that order, this will help in times of difficult decisions. No one said pandemic, epidemic, or crisis management decisions are easy. This is only one example of the pressures that may be on a pandemic and spin-off sub-crisis management teams. Consider the expectations, judgments, and fears of members of the pandemic management team. These include the leader, the pandemic/crisis management team facilitator, among others.

The expectation of the leader is that he or she will lead the pandemic/crisis management team to a rapid and successful outcome. This expectation will put a lot of pressure on that individual. There is a judgment that he or she can lead the pandemic management team. The fear is that no matter what the pandemic or crises—the 1918 influenzas pandemic, an AIDS pandemic, or the 2019 COVID-19 pandemic—a reality will occur, and the teams must be prepared.

For the facilitator, the expectation is that he or she is knowledgeable of the pandemic or sub-crisis plans, is capable to lead the pandemic management team through the plan, and that the plan being used to manage the pandemic is up to date and has been tested. The fear is that he or she will leave out an important part of the plan in a stressful situation.

The judgment is that the plan facilitator can carry out the plan with the existing staff while handling all the other assigned activities. The bottom line is that members of the pandemic/crises management teams and others involved in a pandemic must move from the emotion of the pandemic/crises to managing the pandemic/crises. The way there is through training to develop confidence to satisfy the pandemic/crises situations and the conducting of pandemic/crises exercises using scenarios and mock drills not only to test the plan but also to test moving from the emotions of the situation to the management of the pandemic/crises.

Behaviors exhibited after the crisis has stabilized

This means that the spikes in the number of cases dropped, hospitalizations are down, and the number of COVID-19 deaths are down. Remember we are talking about stabilization, not the end of the pandemic.

Behaviors after a pandemic / crises situations have stabilized fall into categories. First, there is relief. Second, a sense of reality sets in regarding what has happened and the possible ramifications. Third, there may be a sense of guilt: "Why them, and why not us or me?" Fourth, there will be questioning of "Did I do enough? Should we have done some things sooner, and would it have made a difference?" Second-guessing regarding what was done or not done is not helpful at this point. A critique of the event will be helpful sometime after the pandemic is concluded; however, this may be a month or two after the pandemic when heads are clearer. It also must be remembered that the pandemic/crises are not over. While the pandemic situation is considered stable, there is a great deal of work yet to be done to satisfy the crises that developed because of the pandemic. While the feeling of some confidence may be experienced and the ability of pandemic team members to return to their job functions exists, the pandemic management team members may find themselves back in the pandemic management room should the situation deteriorate or if the pandemic manager sees the need for them to return due to virus variants. The pandemic team members must stay focused on the situation and keep in touch with the pandemic manager and each crisis manager.

CHAPTER 9

The Event (Cluster of Illnesses) and the Criteria to Determine a Pandemic, Epidemic, or Crisis

A cluster of illness could lead to a pandemic, epidemic, and from there to many spin-off sub-crises taking many forms. These include:

- One or more people are dead from an illness, and there is no understanding of its origin.
- Multiple illnesses and deaths occurring in various parts of the country and/or world.
- Post the identification of a cluster of illnesses, reporters and television cameras are at locations where the illness clusters are taking place.
- Large numbers of people are calling in sick and not coming to work at many areas of the country and world.
- Commercial products are thought to be responsible for illness.
- Many others.

A cluster of illnesses, should it occur, is usually communicated by telephone to a local health department. If the cluster in the United States is judged to possibly be significant, the state health department contacts the United States Center for Disease Control and shares the information. Considering the potential for the illness cluster becoming an epidemic or even a pandemic, the information is shared with the World Health Organization.

The World Health Organization is the agency responsible for declaring a pandemic. They monitor disease activity on a global scale through a large network of centers of surveillance that are in countries worldwide and has a preparation plan that consist of six phases. These phases are known as alert phases. Phase 1 is the lowest level of alert, and phase 6 is considered the pandemic phase.

The World Health Organization pandemic phases were developed in 1999 and revised in 2005. The phases are applicable to the entire world and provide a global framework for pandemic preparedness and response planning. In this version, the World Health Organization has retained the use of a six-phase approach for easy incorporation of new recommendations and approaches into existing national preparedness and response plans. The grouping and description of pandemic phases have been revised to make them easier to understand, more precise, and based upon observable phenomena. Phases 1 to 3 correlate with preparedness, including capacity development and response planning activities while phases 4 to 6 clearly signal the need for response and mitigation efforts; furthermore, periods after the first pandemic wave are elaborated to facilitate post-pandemic recovery activities.

The following are the World Health Organizations pandemic phases:

In phase 1, there are no viruses circulating among animals that have been reported to cause infections in humans.

In phase 2, an animal influenza virus circulating among domesticated or wild animals is known to have caused infection in humans and is therefore considered a potential pandemic threat. An influenza's virus is identified here, but the protocol is appropriate for all viruses, including COVID-19.

In phase 3, one would consider a human animal influenza reassortment virus has caused sporadic cases or small clusters of disease in people but has not resulted in human-to-human transmission sufficient to sustain community level outbreaks. Limited human-to-human transition may occur under some circumstances, for example, when there is close contact between an infected person and an unprotected caregiver. However, limited transmission under such restricted circumstances does not indicate that the virus has gained a level of transmissibility among humans necessary to cause a pandemic.

Phase 4 is characterized by verified human-to-human transmission of an animal or human animal influenza reassortment virus able to cause community level outbreaks. The ability to cause substantial disease outbreaks in the community marks a significant upward shift and the risk of a pandemic. Any country that suspects or has verified such an event should urgently consult with the World Health Organization so that the situation can be jointly assessed and a decision taken by the affected country if implementation of a rapid pandemic containment operation is warranted. This would indicate a significant increase in risk of a pandemic but does not necessarily mean that a pandemic is a foregone conclusion.

Phase 5 is characterized by human-to-human spread of the virus into at least two countries in one World Health Organization region. These regions are identified as (1) Greenland, Europe, Russia/Siberia, and northeast Africa; (2) north and south America; and (3) China and Australia. While most countries will not be affected at this stage, the declaration of phase 5 is a strong signal that a pandemic is eminent and that the time to finalize the organization, communication, and implementation of the plan mitigation measures is short.

Phase 6, the pandemic phase, is characterized by community level outbreaks and at least one other country in a different World Health Organization region in addition to the criteria defined in phase 5. Designation of this phase will indicate that a global pandemic is underway.

Summary of the six World Health Organization (WHO) phases:

1. No viruses among animals
2. Viruses circulating among animals
3. Small clusters of disease in people
4. Verified human to human transmission

5. Human-to-human spread into at least two countries in one WHO region
6. Community outbreaks in at least one other country in a different WHO region in addition to the criteria in phase 5

The 6 WHO World Regions Used to Declare a Global Pandemic

- African Region
- Region of the Americas
- Southeast Region
- European Region
- Eastern Mediterranean
- Western Pacific Region

During the post-peak, pandemic disease levels in most countries with adequate surveillance will have dropped below peak observed levels. The post-peak signifies that pandemic activity appears to be decreasing; however, it is uncertain if additional waves will occur, and countries will need to be prepared for a second wave. Previous pandemics have been characterized by waves of activities spread over months. Once the level of disease activity drops, a critical communications task will be to balance this information with the possibility of another wave. Pandemic waves can be separated by months, and an immediate at ease signal may be premature. COVID-19 had several waves in the United States.

In the post-pandemic, influenza disease activity will have returned to normal levels seen for seasonal influenza. It is expected that the pandemic virus will behave as a seasonal Influenza A virus. While the example addresses influenzas, it is applicable for all viruses, including COVID-19. At this stage, it is important to maintain surveillance and update pandemic preparedness and response plans accordingly. An intensive phase of recovery and evaluation may be required.

This phased approach is intended to help countries and other stakeholders when certain situations will require decisions and decide at which point actions should be implemented. As in the 2005 guidance, each of the phases applies worldwide once announced. However, individual countries will be affected at different times. In addition to the globally announced pandemic phase, countries may want to make further national distinctions based upon their specific situations. For example: countries may wish to consider whether the potential pandemic virus is causing disease within their own borders and neighboring countries or in countries in close proximity.

Periodically, the phases are updated. It is important to stress that the phases were not developed as an epidemiological prediction but provide guidance to countries on the implementation of activities. While later phases may loosely correlate with increasing levels of pandemic risk, this risk in the first three phases is simply unknown. It is therefore possible to have a situation which poses an increase pandemic risk but not result in a pandemic.

Alternatively, although global influenza surveillance and monitoring systems are much improved, it is also possible that the first outbreaks of a pandemic will not be detected or recognized. For example: if symptoms are mild and not very specific, an influenza virus with the pandemic potential maintain relatively widespread circulation before being detected; thus, the world phase may jump from phase 3 to phase 5 or 6. If the rapid containment operations are successful, phase 4 may revert to phase 3.

When making a change to the global phases, the World Health Organization will carefully consider whether the criteria for a new phase have been met. This decision will be based upon all credible information from global surveillance and from other organizations. Please note that the above refers to an influenza pandemic like the one that occurred in 1918–1919. These same phases are appropriate in addressing the COVID-19 pandemic.

The 2009 pandemic phases:

- Planning tools
- are simpler, more precise, based on verifiable phenomena
- will be declared in accordance with the IHR[2] (2005)

---

[2]   IHR—the International Health Regulations.

- Only loosely corresponds to pandemic risk.
- Identify sustained human-to-human transmission has a key event.
- Better distinguish between time for a preparedness and response; and
- include the post-peak and post-pandemic for recovery activities.

The new phases are not:

- designed to predict what will happen during a pandemic; and
- always going to proceed in numerical order.

IHR represents an agreement between 196 countries, including all WHO member states, to work together for global health security. In the United States, CDC works with state and local reporting and response networks to receive information at the federal level and then respond to events of concern at the local and federal levels.

The World Health Organization has developed pandemic phase descriptions and main actions by phase. These actions are broken into five categories. They are planning and coordination, situation monitoring and assessment, communications, reducing the spread of disease, and continuity of health-care provision.

The planning and coordination actions for phases 1, 2, and 3 are as follows: develop, exercise, and periodically revise national pandemic preparedness and response plans. The situation monitoring and assessment actions for phases 1, 2, and 3 are as follows: develop a robust national surveillance system in collaboration with national animal health authorities and other relevant sectors. The communications actions for phases 1, 2, and 3 states, complete communications planning and initiate communications activities to communicate real and potential risks. The reducing the spread of disease actions for phases 1, 2, and 3 states, promote beneficial behaviors and individuals for self-protection. Plan for the use of pharmaceuticals and vaccines. The continuity of health-care provision actions for phases 1, 2, and 3 states, prepare the health system to scale up.

The actions identified by the World Health Organization for phase 4 for planning and coordination state direct and coordinate rapid pandemic containment activities and collaboration with the World Health Organization to limit or delay the spread of infection. Concerning situation monitoring and assessment for phase 4, the World Health Organization states, increase surveillance, monitor containment operations, share findings with the World Health Organization and the international community. Regarding communications, the World Health Organization states for phase 4, promote and communicate recommended interventions to prevent and reduce population an individual risk. Regarding reducing the spread of disease for phase 4, the World Health Organization states, implement rapid pandemic containment operations and other activities; collaborate with the World Health Organization and the international community as necessary. Regarding continuity of health-care provision, the World Health Organization states for phase 4, activate contingency plans.

The action plans for phase 5 and 6 under planning and coordination states, provide leadership and coordination to multisectoral resources to mitigate the societal and economic impacts. Under situation monitoring and assessment for phase 5 and 6, the World Health Organization states, actively monitor and assess the evolving pandemic and its impacts and mitigation measures. Concerning communications for phases 5 and 6, the World Health Organization states, continue providing updates to general public and all stakeholders on the state of pandemic and measures to mitigate risk. Main actions for phases 5 and 6 under reducing the spread of disease the World Health Organization states, implement individual, societal, and pharmaceutical measures. The actions under continuity of health-care provision for phases 5 and 6 as stated by the World Health Organization includes implementing contingency plans for all health systems at all levels.

For the post-pandemic

Actions identified by the World Health Organization under planning and coordination state, plan, and coordinate for additional resources and capabilities during possible future waves.

Actions under the post-peak

Identified by the World Health Organization as many actions under situation monitoring and assessment, the World Health Organization states, continue surveillance to detect subsequent waves under the post-peak. Under main actions under communications, the World Health Organization recommends, regularly update the public and other stakeholders on any changes to the status of the pandemic under the post-peak. For actions under reducing the threat of disease by the World Health Organization, it states, evaluate the effectiveness of the measures used to update guidelines, protocols, and algorithms in the post-peak. Under actions concerning continuity of health-care provision, the World Health Organization states, rest, restock resources, revise plans, and rebuild essential services under the post-pandemic. Actions identified by the World Health Organization under planning and coordination states, review lessons learned and share experiences with the international community; replenish resources. Under the post-pandemic for situation monitoring and assessment, it states, evaluate the pandemic characteristics and situation monitoring and assessment tools for the next pandemic and other public health emergencies. Under the post-pandemic main actions and communications, it states, publicly acknowledge contributions of all communities and sectors and communicate the lessons learned; incorporate lessons learned into communication activities and planning for the next major public health crisis. Under the post-pandemic and the actions identified as reducing the spread of disease, the World Health Organization states, conduct a thorough evaluation of all interventions implemented. In the post-pandemic under the section called continuity of health-care provision, the World Health Organization states, evaluate the responses of the health system to the pandemic and share the lessons learned. Obviously, to respond appropriately at not only the World Health Organization level but the state and local levels, there is a need for accurate and timely information being provided. In the United States, the Center for Disease Control has established a system for gathering information/data.

The World Health Organization has a detailed plan using six phases before they decide the world is faced with a global pandemic. This should be satisfying people who believe that the decision of the World Health Organization to identify a pandemic is done on a whim. It is not. However, the quality of communicable disease data received by the Center for Disease Control in the United States and other countries is critical to accuracy; and accuracy of health data is critical to acceptability by the WHO and other agencies gathering and receiving health data so that informed health decisions can be made by the WHO and other health management groups.

It will be necessary to gather information from many countries to manage communicable diseases. For this, a national communicable disease surveillance system has been developed. This will increase the accuracy in the acceptability of data gathered. It will help establish and coordinate the setting of effective policies and procedures and provide appropriate communication infrastructure for data exchange, and defining a communicable disease minimum data set are essential.

Another way of looking at this is to say science and opinions are not the same thing. While science does not speak, it does provide information so that experts in the area can make informed decisions based on the data not a whim. Therefore, when the World Health Organization, using worldwide data that meet their criteria, state, "We have a world pandemic," you can count on its accuracy and correctness.

Reportable diseases are diseases that are of great public health importance. The United States Center for Disease Control and Prevention requires the identification of diseases to be reported when they are diagnosed by doctors and laboratories. The statistics that show how often the disease occurs helps researchers identify disease trends and tracks outbreaks that helps control future outbreaks.

The reportable diseases are divided into several groups. These include mandatory written reporting, mandatory reporting via telephone, report a total number of cases. Examples: chickenpox, influenza, and cancer. Cancer cases are reported to the state cancer registry. Examples of diseases reportable to the CDC include but are not limited to anthrax; botulism; chickenpox; gonorrhea; hepatitis A, B, and C; HIV infection; influenza-related infant deaths; lead elevated blood levels; legionnaires disease; leprosy; malaria; measles; meningitis; mumps; novel influenza; virus infections; pesticide-related illnesses and injuries; the plague; severe acute respiratory syndrome associated coronavirus diseases; COVID-19; smallpox; tuberculosis; yellow fever; cholera; and diphtheria. The information gained from reporting allows the country or state to make informed decisions and laws about activities and the environment, for example, flood handling, immunization programs, insect control, water purification, and more.

Examples of spin-off crises occurring because of the pandemic include but are not limited to loss of job, school closures, where both parents work and one parent will need to leave their job to care for the children, food shortages, material shortages, among others, will need to be addressed. There is a question of who should manage these crises. Regardless of who learns of the crisis' issues, a crisis should be immediately referred to an individual in a leadership role. If this person is out of town, a second-in-command should receive the call and later notify the person in charge. If the call announcing a pandemic is received after business hours and the business is closed or if the call occurs on a weekend or on a holiday, a call list should be available to contact the appropriate management individual. A third-party answering service is not desirable or recommended in reporting an illness cluster that may lead to a pandemic.

It is important to remember that the individual phoning in the information concerning an illness cluster may be upset and have difficulty relating information and not thinking clearly. They may want to quickly provide the information and return to help others. The management person receiving the call should note the location of the illness cluster, the date and time, the name of the person providing the information, and the name of the person receiving the information. The following are examples of questions that should be asked of the person reporting the illness cluster:

What is the nature of the illness cluster, and where possible identify the extent of the cluster? Obtain the phone number of the person providing information should they be cut off or if you need them to provide additional information. What is the name of the hospital where people have been taken? Are the news media covering the situation? If yes, what questions are they asking? Who is answering the questions? And what information has been provided to them? If they are on the scene, who are they? Do they include television crews, newspaper, or magazine press and/or radio? Does the hospital receiving COVID-19 cases need help? This includes medical help. Do they need ambulances for transport to hospitals? And do they need people to help with counseling? The person supplying this information may not know the answers to these questions or may be too upset to supply some of the above information.

Keep in mind the last thing the person reporting an illness cluster wants to do is to a provide the same information many times to different individuals. Ideally, the information should be provided only once, and that information should be shared with those on a need-to-know basis. All the above information and more needs to be provided to effectively respond to the situation. Further, it needs to be determined if a press release or standby statement has been prepared or provided to the media. If no, the standby statement needs, if possible, to be forwarded to a pandemic management team for approval. In some cases, timing interferes with the approval process. This is where the media is waiting for the statement, and any delay in providing information may be viewed as stonewalling. There should be a list of internal experts, in some cases global experts, who could respond first by telephone and, if needed, fly directly to the areas of need. There should also be a list of approved outside consultants and their phone numbers. For example, communications and technical experts so that they may, as needed, be drawn in to aid not only to satisfy a pandemic but also the spin-off crises as well.

There is also a need to have up-to-date phone list to respond to both a pandemic and spin-off crises. As phone lists are put together, it is pretty much guaranteed that, within fifteen minutes of putting the list together, there will be an error. The reason is that someone has changed their cell phone number or their home phone number since the last time the phone list was reviewed and updated. To ensure that all the needed information is gathered, prepare two wallet-sized, laminated cards. One contains the questions to be asked of the those providing information that may possibly lead to the WHO declaring a pandemic, and the other contains the names and phone numbers of those who may be contacted to assist in the pandemic or subsequent crises. The questions identified above should be made part of a wallet-sized trifold that should be easily carried by all people in leadership positions so they may be prepared to ask questions of the person reporting the information concerning a pandemic or spin-off crises. An example of the trifold questions may be found in appendix 1. Similarly, a wallet-sized trifold card should be carried by leadership with the names and phone numbers of resources that are available to help at times of pandemic or spin-off crises.

An example of this resource list may be found in appendix 2. Timing is important when a pandemic occurs. The correct decisions, with the right information, can make the difference between the pandemic, resulting in none to multiple spin-off crises. There should also be a separate up-to-date list of approved outside consultants available to the pandemic management and, as needed, crises management teams.

An individual must be assigned the responsibility of updating the phone list, the internal experts list, and the list of outside consultants. It is suggested that these lists are updated once per quarter; however, computer programs may be established so that individuals may report changes to their respective phone numbers. One person is needed to oversee the lists for accuracy and whether or not they are up to date. The last thing that is needed is to make a phone call to an expert or consultant who is needed to help in a pandemic or crises and finding that they have changed their phone number, and therefore, you are not able to reach them. While quarterly updates are not perfect, they are a step in the right direction.

With the pandemic or subsequent crises identified, there is a need for emotion-free internal discussions to better understand what has occurred so informed decisions may be made. For those not able to attend the pandemic management team meeting, a conference call should take place to discuss and determine a level of action. There needs to be clear definition of terms and criteria to guide the pandemic management team such that the team will be able to make informed decisions.

Possible criteria to assist the team could include the following:

- Identify the number of people that are dead or ill because of the pandemic and subsequent crises
- The impact on the community
- The expected loss of jobs
- The expected loss of business, actual or anticipated
- The loss homes due to default on mortgages
- Other

The group making the recommendation can suggest that, because of the pandemic, several crises be declared. When the pandemic management team is unsure as to what to recommend, it is important to err on the side of caution; be open with the leadership and explain why the group is or is not convinced a pandemic should be declared in our country regardless of what the World Health Organization has declared.

The leader must listen carefully, ask questions, and take a decision as to whether to declare a pandemic or spin-off crises in their country, keeping in mind that the World Health Organization may have already declared a global pandemic. There may be a need to acquire additional information before a decision is taken.

Matrices may be used to help determine if a pandemic exists. This will be discussed in chapter 13. A pandemic is decided by the World Health Organization, and the impact of the pandemic on each country, such as the United States, must be evaluated by a country pandemic management team. The declaration of a country pandemic and subsequent spin-off crises are expensive. It costs money because the action will involve the commitment of people resources, time, and money. For this reason, only the pandemic management team will make the decision. This approach eliminates a country pandemic declaration without the use of criteria identifying why the circumstances represent or do not represent a pandemic situation, and it prevents dilution of efforts.

With the WHO declaration of a pandemic, it will be difficult to impossible not to declare a country pandemic. Having said this, the WHO declared HIV as a pandemic. While it was a global pandemic, its direct effect on US citizens was and is far less than what we are experiencing with the COVID-19 virus. If a country pandemic is declared, the effort to address the pandemic and spin-off crises will take away resources that may be needed elsewhere but will also be needed to address the pandemic.

If a pandemic or crisis is declared, work begins immediately with the group responsible for communication. This group begins preparing a statement and sound bites to be given to the media. The pandemic management team should have a standby statement. This statement should be prepared as a joint effort of the communications group and the pandemic management team. When appropriate and approved, it should be issued to the media.

What is a standby statement, and when and how is it used?
Standby statement
Get ready—prepare a standby statement
Get set—approve the standby statement
Go—issue the standby statement

A pandemic or a spin-off crisis has occurred, and the media are on the scene. The media want to take pictures and conduct interviews. The media will write a story about the pandemic or spin-off crises, whether or not they receive information; therefore, it is best to have something prepared to provide information to the media. This standby statement is provided to the media to supply basic information and thereby hopefully prevent negative stories. The standby statement should provide factual information. At no time should it contain opinion. This is so important that I shall say it again. At no time should a standby statement include opinion. The written standby statement has blank areas for information about the pandemic or spin-off crises, which will be added prior to providing the statement to the media. The scope of the standby statement should be broad, but at the same time, it must be focused on the pandemic and/or spin-off crises because of the pandemic. For example: "Today [give date] the World Health Organization declared a global pandemic in response to the COVID-19 virus. We are responding to this pandemic and possible subsequent sub-crisis, should they occur, with every resource at our disposal. We have requested outside assistance [state only if true] from experts. We have confirmed [give a number as appropriate] people dead and hospitalized [give number as appropriate]. Our Center for Disease Control (CDC) will oversee the acquiring of up to date and accurate statistics on cases." This section of the standby statement identifies dead and hospitalized and must be accompanied by a statement of sympathy and empathy for those that have died and their families. The standby statement should be read to the media by pandemic management team spokesperson at the first gathering with the media. This is not a press conference, only a meeting to provide up-to-date information. Once the completed standby statement has been read and handed out to the media, the spokesperson should then say, "I know you have questions, but I only have time for three since I am needed to continue to work with the pandemic management team. I thank you for your understanding and attention."

CDC Headquarters, Washington, D. C.

After three questions are asked and answered, the spokesperson is to excuse him or herself to return to the pandemic management team. The answers provided by the spokesperson must be factual and not opinion. If he or she does not know the answer to the question, he or she should answer, "I do not know the answer to that question but shall try to get the answer and provide it at the press conference." If messages have been created by the communications sub-team and approved by the pandemic or spin-off crisis management teams, they should be presented by the spokesperson without taking a breath after each question is answered. However, if they have not been developed and approved at this point in time, the messages must be ready at the time of the press conference. The standby statement is designed to provide measured and accurate information.

Without the standby statement and a spokesperson, there may be an attempt to conduct interviews with other people not trained to be interviewed by the media. It is suggested that if anyone other than the spokesperson is approached for an interview by the press or television media, the interviewer should be referred to the spokesperson for information. Further, they should explain that the situation would be best served if they could leave and help those who are affected by the pandemic. The standby statement should close with an estimated time until the next briefing, the press conference. This could be hours, but not more than twenty-four, and the words "standby statement" should not be on the document. An example of a crisis standby statement may be found in appendix 3.

It is important to have people who have undergone training on how to be interviewed by the media. The training should be considered critical, and without it, opinions and comments to the media may become problematic. This training is highly recommended.

In summary

The standby statement should be read by the leadership's spokesperson, and copies should be handed out after only three media questions are answered by the spokesperson. Keep in mind, this is not a press conference; it is a short news briefing. The standby statement is designed to give the media something to report while they wait for the press conference. It is very important that, once announced, the timing for the press conference be kept. If your spokesperson says that the press conference will be in four hours, it must take place within four hours, or creditability will be lost.

Once a pandemic or spin-off crisis has been declared and the standby statement drafting assigned, the pandemic management team must be immediately called to convene if they have not already met. The staff of the communications group should contact members of the pandemic management team, and contact should be made by telephone, by e-mail, or both. The pandemic management team is provided the following information:

- a brief description of the pandemic from the WHO
- the date, time, and location where the pandemic management team will meet
- a phone number for those pandemic management team members who are out of town and cannot come to the location where the team is gathering

If videoconferencing will be used, all appropriate information should be provided. Team members or their designees are expected to be present or be on the telephone at the designated date and time. This includes people on travel and in other countries around the world. There may be more than one possible location for the team to meet, so team members need to be aware of the meeting location.

The declaration of a pandemic and spin-off crises can get complicated. With a global pandemic and without coordination, the pandemic may be managed in different ways in different parts of the world. For this reason, there must be coordination globally, and it must be ongoing. For example: closed boarders will present issues of travel and trade. With a pandemic announced by the WHO, many actions will be taken by each country involved, and procedures and protocols should be in place and agreed globally to limit confusion and wasted time.

CHAPTER 10

———

The Nine Stages of Managing a Pandemic

Unlike a crisis such as an explosion or fire, a worldwide pandemic is first addressed by the World Health Organization (WHO). Therefore, the first four stages described below are addressed by WHO after which each individual country takes actions, stages 5 through 9, to manage the pandemic. While the WHO have provided comments on addressing pandemic, cultures vary from country to country. As a result, while WHO provides guidance, the management of the pandemic should be on a country-by-country basis.

While experiencing and working through the many emotions associated with a pandemic, it is important to understand and focus on the nine stages of pandemic management. Understanding the stages can be critical in creating and implementing the appropriate emergency or pandemic response. Also, the value of separating a pandemic into stages is to understand where you are in addressing the pandemic, where you have been relative to the pandemic, and what are the next steps. These stages include stage 1, the event. In this case, the event is the WHO identification of a novel virus; stage 2, the rapid gathering of information to determine the seriousness of the event; stage 3, the WHO decision whether or not a pandemic situation exists or could exist in the foreseeable future, the WHO declaration that a pandemic exists; stage 4, if a pandemic is declared, working to minimize the impact of the pandemic and addressing the crises that are spin-offs of the pandemic without being sidetracked and taking

127

your eye off the pandemic. Stabilizing the pandemic, assessment of the damage and community impact, remediation, recovery, return to normal or as close to normal as possible, and closure, the bringing of the pandemic to an end—the latter are stages 5 through 9.

Stage 1: the event

Pandemics will vary as to their severity and probability of occurrence. They would, of course, be health related. The discovery may be in a laboratory where a new of virus is discovered. It could be because of a mutation of a known virus, or it could be the finding of a new virus in bats or animals. With the event, the finding a new virus, usually comes concern, confusion, fear, and often and depending on the ability to quickly learn about the virus, frustration.

Questions evolve as to:

- Are we aware of any known virus with similar structure and characteristics?
- What is the potential for this virus causing a pandemic?
- Where are the resources to go forward in evaluating this virus?
- Many others

The answers to these and other questions are not easy and, in some cases, almost impossible to determine. There is a potential that perhaps hundreds of thousands or millions of lives may be lost if a pandemic occurs because of this virus. More information is needed, and it is needed as soon as possible.

Stage 2: the rapid gathering of information by the WHO to determine the seriousness of a potential pandemic

With post-event confusion and the setting in of reality as to what occurred, there is a need for WHO to gather factual information concerning the event. For example: does the location that discovered the virus need help? And if so, what type of help? Will people die who are infected by this virus? How can it be transmitted? Is transmission animal to people, people to people? What crises will develop from the pandemic because of the virus? And who and what will be affected, and when? How will the information about the virus be communicated? It needs to be done without creating a panic. It needs to be informative and without jargon. Who will prepare the messages for the media? Who will review them? And will they need to be approved?

It needs to be remembered that, with an accident, if one person dies, the number of people that will be affected by the dead are many: spouses, families, coworkers, and the list goes on. It is not impossible that if a million people experience a loss due to a pandemic, ten to twenty million others are impacted by the million deaths. As of now, more than 5 million people have died worldwide from the COVID-19 pandemic from January 2019 through mid-November 2021, and some 254,000,000 million have come down with the virus. This means that 2 percent of those who came down with the virus have died. If we use the same relationship that ten to twenty people are aware or personally know someone who has had COVID-19, between 2.5 billion and 4.9. billion people have been in some way been impacted by the virus. In other words, somewhere between 25 percent to almost 50 percent of the people in the world have in some way been affected either physically or emotionally by this virus.

This is to say nothing of people out of work, no money, and in need of food. The point here is that this is a very serious situation that needs to be addressed globally by knowledgeable and experienced people. It is more than likely that you, a loved one, a relative, or friend know someone or you know someone who knows someone that has been impacted by this virus. In the area where you live, how many deaths have occurred? Are there people out of work due to the closing of the location where they worked? Is the hospital in your area at or near capacity? Are there enough first responders and medical personal available? Are you receiving nurses and doctors from locations outside your area, and are they on the way to a hospital stretched to its limits and in need of help? Is the media involved at the hospital, or are they present in areas heavily impacted by the virus? Who is speaking to the media? Has local government issued a standby statement? If not, who is prepared to issue a press release or a standby statement to the media? How many crises are developing in your area because of the pandemic?

Stage 3: the decision regarding the existence of a pandemic

The decision by the WHO as to whether to declare a pandemic cannot be taken lightly. The declaration will affect almost everyone in the world. It will cause businesses to close, people to lose jobs, bills not to be paid, and people will lose their homes, among other devastating impacts. Regardless, not declaring a pandemic and taking the necessary action to address it, millions, perhaps billions, of people will be infected by the virus with a very high percentage of them dying. Therefore, the decision cannot be based on a whim or an idea. It must be based on facts, documented facts. The costs associated with the decision to declare a pandemic will be great. However, it must be remembered, save lives first. The economy, property, profits all come after saving and protecting lives. It must also be remembered that crises involving business closings, loss of jobs, product shortages including food, and much more will follow. The decision on declaring a pandemic must be an informed decision and is to be made after deliberation of a gathering of people experienced and knowledgeable of the COVID-19 virus and action to be taken once a pandemic is declared.

Stage 4: the announcement of the decision of the existence of a pandemic comes from the leader of the World Health Organization (WHO)

The WHO declared the COVID-19 impact a pandemic on March 11, 2020. The WHO is responsible for monitoring issues around the world that may develop into epidemics or pandemics. I want to share with you the director general's opening remarks when the COVID-19 pandemic was announced by the World Health Organization.

In the past two weeks, the number of cases of
Covid 19 outside of China has increased 13-fold

and the number of affected countries has tripled. There are now more than 118,000 cases and 114 countries, and 4,291 people have lost their lives. Thousands more are fighting for their lives in hospitals. In the days and weeks ahead, we expect to see the number of cases, the number of deaths and the number of affected countries climb even higher. The World Health Organization has been assessing this outbreak around the clock and we are deeply concerned both by the alarming levels of spread and severity and by the alarming levels of inaction. We have therefore made the assessment that Covid 19 can be characterized as a pandemic. Pandemic is not a word to use lightly or carelessly. It is a word that, if misused, can cause unreasonable fear, or unjustified acceptance that the fight is over, leading to unnecessary suffering and death. Describing the situation as a pandemic does not change the World Health Organization assessment of the threat posed by this virus. It does not change what the World Health Organization is doing, and it does not change what countries should do. We have never seen a pandemic spark by coronavirus. This is the first pandemic caused by the coronavirus.

Later in the speech, the director general said, "We know that these measures are taking a heavy toll on societies and economics just as they did in China. All countries must strike a fine balance between protecting health, minimizing economic and social disruption, and respecting human rights." Near the end of the speech, the director general said, "First we must prepare and be ready. Second, detect, protect, and treat. Third reduce transmission. Fourth, innovate and learn." Near the very end of the speech, he added, "Prevention, preparedness, public health, political leadership and most of all people."

As key elements for a path forward, I included segments of the speech because it is clear the World Health Organization recognized the fact that there would be crises other than health that came out from the presence of this virus. They gave guidance in how to prepare for what was coming. In making an informed decision concerning existence of a worldwide pandemic, the following are examples of information that need to be gathered.

In gathering information, it was learned that the COVID-19 virus had the possibility of affecting 10 percent or more of the world's population. In March 2020 the CDC learned that COVID-19 was a disease occurring over a wide geographical area and affecting an exceptionally high proportion of the population. It has been determined that the first COVID-19 case was identified as being from Wuhan, China, in December 2019. Starting in Wuhan, China, it quickly became an epidemic because the disease then spread across several countries and affected the large number of people. An epidemic is a disease that affects many people within the community, population, or region. A pandemic is an epidemic that spread over multiple countries or continents. A pandemic affects more people and takes more lives than an epidemic. The World Health Organization declared COVID-19 to be a pandemic when it became clear that the impacts were severe, and that it was spreading quickly over a wide area. An endemic is something that belongs to a particular people or country. The WHO began assessing this outbreak around the clock and were deeply concerned both by the alarming levels of spread and severity and the alarming levels of inaction. This book does not spend much time addressing an epidemic since it may be considered a "local" pandemic, country or region driven.

An epidemic is a disease that affects many people within a community, population, or region.

A pandemic is an epidemic that spread over multiple countries or continents.

On January 21, 2020, a Washington state resident became the first person in the United States with a confirmed case of the 2019 novel coronavirus, having returned from Wuhan on January 15, thanks to overnight polymerize chain reaction testing. On March 5, 2020, the National Institute for Communicable Diseases confirmed that a suspected case of COVID-19 had tested positive. The patient was a thirty-eight-year-old male who travel to Italy with his wife. Cases began to be reported in Africa, reaching the continent through travelers returning from hotspots in Asia, Europe, and the United States. Africa's first COVID-19 case was recorded in Egypt on February 14. Since then, more than fifty-two countries have identified COVID-19 cases.

It is interesting that the coronaviruses are an extremely common cause of colds and other upper respiratory infections. The World Health Organization was aware that their decision to declare a pandemic would cause many different problems throughout the world. This would include business shutdowns, job losses, and many more hardships. However, the World Health Organization must focus on lives first. They should not and did not consider any hardships that may result from crises that came about because of the pandemic. This was the right and only decision for the World Health Organization.

It was determined that the virus was moving from person to person. There are several ways this can happen: droplets or aerosols. This is the most common transmission. When an infected person coughs, sneezes, or talks, droplets of tiny particles called aerosols carry the virus into the air from their nose or mouth.

The World Health Organization compared the COVID-19 pandemic with other epidemics and pandemics. For example: in 1918 the influenza pandemic was the most severe pandemic in recent history. It was caused by an H1N1 virus with genes of avian origin. Although there is not universal consensus regarding where the virus originated, it spread worldwide during 1918 and 1919. Other epidemics and pandemics that were reviewed by the world health organization included:

Epidemic/pandemics	Deaths	Date
Black death	75–200 million	1346–1353
Spanish flu	17–100 million	1918–1920
Plague of Justinian	15–100 million	541–549
HIV/AIDS	35+ million	1981–present

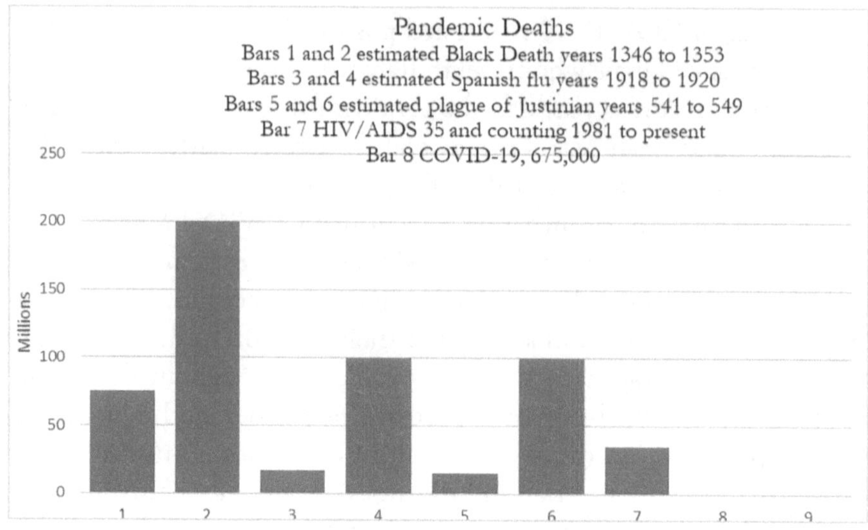

Bars identifying pandemic deaths.

The World Health Organization explained how to control the spread of the virus. They suggested the following:

- Avoid close contact with people who are sick.
- Minimizing touching your eyes, nose, and mouth.
- Stay home when you are sick.
- Cover your cough or sneeze with a tissue and throw the tissue in the trash.
- The time from exposure to symptom onset (known as the incubation period) is thought to be two to fourteen days,
- The symptoms typically appear within four to five days after exposure.
- A person with COVID-19 may be contagious forty-eight hours before starting to experience symptoms.
- The symptoms may include fever, fatigue, dry cough, loss of appetite, body aches, shortness of breath, and mucus or phlegm.
- Walking past someone in the street or having a jogger run by you means you are close together for a few seconds, and

most fleeting encounters are highly unlikely to be long enough for the virus to reach you.

- Washing your hands with soap and water for at least twenty seconds, especially after you have been in a public place.
- After blowing your nose, coughing, or sneezing, throw away the tissue that was used.
- It is especially important to wash your hands before eating or preparing food and before touching your face.
- Other ways of controlling the disease in addition to washing your hands is the wearing of a mask or a cloth face covering. The face covering blocks entrance of the virus through the nose and mouth, and if a person has COVID-19, it prevents COVID-19 spreading great distances.
- Maintain social distancing. Six feet is recommended because of the high molecular weight of the virus. This results in the virus going to the ground unless you have projectile sneezing, projectile coughing, singing, yelling, and the like, which will carry the virus greater distances.
- It is important to clean surfaces. While the virus is not considered alive, it cannot reproduce. The disinfectants will interfere with the chemistry of the virus and should be used.
- The World Health Organization recommends stopping sharing of food, drink, clothing, and the like.

From the above, you cannot understate the value of the WHO during pandemic situations.

The longest enduring pandemic disease outbreak is the seventh cholera pandemic, which originated in Indonesia and began to spread widely in 1961. As of 2020, some fifty-nine years later, this pandemic is still ongoing, and in fact, it effects an estimated 3,000,000 to 5,000,000 people annually.

Research suggests that COVID-19 is not active for long on clothing compared to hard surfaces and exposing the virus to heat may shorten its activity. A published study found that at room temperature, COVID-19 was detectable on fabric for up to two days, compared to seven days for plastic and metal.

Two vaccines developed in the United States have demonstrated a high degree of efficiency. They have been found to be 94.5 percent effective. At this point, it is not clear how long the shots will last. Some believe forever; others believe there will be a need for a booster. A booster is now recommended six months after the second shot. It has been recommended that those over sixty-five years old and those with compromised immune systems consider getting a booster shot prior to eight months after the second shot. If the first responders and medical personnel who first received a vaccine begin to develop signs and symptoms of the illness having been vaccinated, indicating breakthrough infections, a booster may be necessary. It is important to keep in mind that nothing is guaranteed. Many things will affect whether a person exposed to COVID-19 will become sick or not. These include safety measures and your personal immune system.

As with the WHO, those who will manage the pandemic at a world, country, or regional level must, first and foremost, focus on saving lives while working toward bringing the pandemic and subsequent resulting crises to stability. The effort to reach stability may take many forms; for example, it may involve the identification and directing of resources. These resources can be, depending on the stage of the pandemic, health experts, virologists, people knowledgeable in media relations, remediation and recovery teams, and teams to address the emotional impact on people losing their jobs, homes, and the lives of their relatives and friends in communities, regions, territories, or country. In addition, it is important to contact stakeholders such as the local community and its leaders, the local police and fire departments, regulatory agencies, hospitals, medical communities, to name only a few, and identify preparations for a pandemic and subsequent crisis.

Stage 5: stabilization of the event

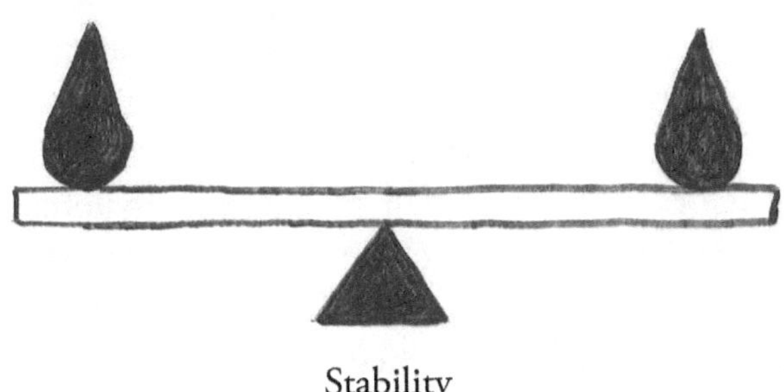

Stability

Stability means the pandemic is being managed. The number of deaths is on a steady downward trend. Virus variants are under control. Hospitalizations from the virus are minimal if at all. Businesses are reopening. Schools are beginning in school classes. Unemployment is down and approaching pre-pandemic numbers. Home foreclosures are at or below pre-pandemic numbers, and confidence is returning to the country. However, the fact that led to the pandemic and subsequent crisis situations is judged to be stable does not mean the pandemic and subsequent crises are over, and the work ahead may be long and difficult. Flare-ups and new waves of the disease have and continue to be experienced. The pandemic management team, which may take a step back once stability is reached, should assign a pandemic manager and crisis manager for each crisis management team to address crisis areas to be managed post-stabilization if they have not already been established.

These crisis management teams should address businesses that closed during the pandemic and may not open again, for example: a small vacuum-cleaner company that closed during pandemic because they could not get parts for products. The parts were made in Asia, and the Asian company that supplies their parts is not open since that country does not have enough vaccine to immunize their people so they can return to work to make the parts needed to make the vacuum cleaners. This means that the employees at the vacuum-cleaner company cannot return to work. This is a problem for many companies in the United States. Other countries are far behind in vaccinations, and people in this country are affected by closures in other countries that directly affect employees in this country. The emotional stress for employees in this country who cannot return to work because of the pandemic affecting people in other countries is a real issue and one that is difficult to satisfy.

Other issues in need of being addressed include, but are not limited to, legal issues, supply-chain issues, regulatory issues, remediation, recovery, and closure issues, to name only a few. Closure, in this case, means to have learned from the pandemic and move forward by reopening the business or starting a new business. The pandemic manager or crisis managers of the crisis teams may call back the pandemic management team at any point where it is judged appropriate.

Stage 6: assessment of the damage and community impact post-stabilization

Post-stabilization. There is a need to understand the impact of the pandemic. This includes the emotional impact, health impacts, the business impact, and future costs associated with remediation and recovery. Depending on the level of the impact, the emotional issues that the surviving employees may face include the loss of the life of a relative or close friend because of the pandemic or subsequent crises. For example: the loss of life at a business facility due to the pandemic will affect not only other employees but also the family and friends of those who died. The families and friends of contractors and visitors who may have been on-site during the pandemic will also be impacted.

The physical health and emotional impact from the pandemic and subsequent crises may result in long days for recovery and rehabilitation; and the physical and emotional pain for those with preexisting conditions may take weeks, months, or perhaps years to heal. Again, depending on the extent of the pandemic, business closures and loss of jobs and incomes may take months to fully recover. The business impact could also be severe in that the customer base can shift to other suppliers if there is a fear that a product supply crisis will result in a long-time or permanent disruption in the supply chain for materials customers need to make products for their customers. This is to say nothing of the emotional and educational impact from school closures. While there was online teaching, many students did not have access to computers.

Stage 7: remediation and cleanup

Unlike typical crises caused by explosions, fires, and the like, remediation for a pandemic is very different. Remediation and cleanup of a pandemic refers to disinfecting hospital rooms and cleaning up of equipment such as instruments, hospital gowns, tubing, gloves, masks, treatment drugs, and other materials associated with caring for people with COVID-19 in isolation and in hospitals. This biological waste is addressed as hazardous waste under United States law.

If, however, we look at the remediation and cleanup of the many spin-off crises from the pandemic, this would include equipment that has been shut down for a year or longer due to layoffs that may not start up due to lengthy inactivity or repairs to this equipment so that it may be returned to service. This will take time and cost money, and people who work in manufacturing will remain out of work. Another area where remediation and recovery are necessary include failure to pay mortgages. Here remediation and cleanup are more of the creation of a scheme to permit the people to pay their mortgage with dignity, and without a stigma of failure. Lack of food also needs to be addressed. For people that have lost their jobs do not have money to buy food. Food banks need to be established and filled with food and a mechanism for distribution created or expanded.

It is difficult for many to understand that we are in a global economy. Products/materials and the like are made and traded globally. Therefore, the crisis of lost jobs in one part of the world affects products in another part of the world. The remediation and cleanup of this situation will be complicated. For example: When do you discard the product waiting for the needed part not being made in another part of the world? Do you discard your product where you are waiting for parts in favor of making a product where you can acquire parts, or do you clean up and produce a product where you have a limited supply of all the parts to make the product? Obviously, remediation and cleanup after a pandemic and subsequent crisis is not easy.

Stage 8: recovery—the returning to normal or as close to normal as possible

Recovery will often take time, possibly years. Recovery may be defined as the condition that existed just prior to the pandemic and subsequent crises; however, it is unlikely that the condition just prior to the pandemic will ever be achieved. This is especially the case if there have been excessive deaths such as was and is the case with COVID-19. The best that can be hoped for is to bring about a condition as close as possible to the condition that existed just prior to the pandemic. Keep in mind that over the time of the COVID-19 pandemic, much has changed. People have worked from home and been productive and, as a result, may not choose to return to an office. Children have lost time in school and the important social experiences. It will not be easy.

Critical to the entire effort is what has been learned because of the pandemic and the subsequent crises that followed and how these learnings may be put in place to prevent such an occurrence from happening in the future. While the pandemic management plan and crises management plans that were implemented during the pandemic will be updated because of what was learned, much more is needed. A detailed study of the conditions that led to the pandemic must be carried out, and learnings from this study must be shared and, as appropriate, implemented across the world.

Stage 9: closure—to bring the matter to a close

This is not easy to do and, in some cases, almost impossible. It is almost impossible because of the large number of lives lost to COVID-19. However, what was learned from the pandemic and how it was addressed in bringing it to stability, remediation, and recovery are a part of closure. Further, as part of closure, it is important to ensure that a pandemic does not occur again. This may be difficult, but we must try.

Summary of the nine stages of a pandemic and subsequent crises

- Stage 1: the event; the WHO reviews the possibility of a global pandemic.
- Stage 2: the rapid gathering of information by the WHO to determine the seriousness of a potential pandemic.
- Stage 3: the WHO considers the decision regarding the existence of a pandemic.
- Stage 4: a pandemic is declared by the WHO, and the effort begins to minimize its impact and bring about stability.
- Stage 5: stabilization of the event.
- Stage 6: assessment of the damage and community impact post-stabilization.
- Stage 7: remediation and cleanup.
- Stage 8: recovery—the returning to normal or as close to normal as possible.
- Stage 9: closure—to bring the matter to a close.

CHAPTER 11

Developing or Improving Plans to Address the Pandemic and Spin-Off Crises

Developing and improving "pandemic or crises management plans"

A pandemic management plan is designed to instruct the pandemic management team members on how to respond, not react, during a pandemic situation. The plan, with some minor changes, can be used to manage crises that are spin-offs from the pandemic. The plan should include a distribution list, a place for signatures of those who reviewed and approved the plans, the respective plan owner, and those who approved the plans, and a revision history. The table of contents should include a plan purpose, a scope, references, definitions, responsibilities, procedures, records, the identification of who has been trained in implementation of the plan, appendices, and a document management section. The document should be a controlled document, where signatures are required as approval by all senior staff who review the document. This review and sign-off protocol is followed for each modification to the plan and is especially important when there are changes in management. For this reason, it is not recommended that people's names, only positions, be used in identifying who will sign off on the document. The person signing and approving the document signs their name as holding one of the positions identified in the pandemic management and crises plans. A controlled document following these steps will prevent unauthorized changes to the document.

A pandemic and crisis management plan template may be found in appendix 4.

Please review the template for how to carry out the pandemic or crisis management plans. Each time the plan is modified and signed off, the date of the signing is noted in the plan, which ensures that the plan is reviewed and updated on a regular basis. The plan should be reviewed and modified after each pandemic situation, after each pandemic exercise, or at a minimum of at least once every three years. It is important to identify those who will be holding copies of the plan since not everyone will receive the plan. As a controlled document, the plan can only be amended by the document owner. In most cases, this is the senior individual responsible for among other things a health program.

Plan purpose. The purpose of the plan is identified in the first paragraph of the document. The purpose is typically identified as providing leadership, guidance, and support for rapidly responding to a pandemic or spin off crises in an efficient and effective manner.

The scope of the plan. The scope of any pandemic or spin-off crisis management plans needs to be very broad based and usually includes possible situations such as a pandemic health situation and possible spin-off crises that result such as unemployment, business shutdowns, and lack of product supply to name only three.

One question that arises regarding pandemic or crises management plans is, how large of a document should they be? Some plans are five hundred pages that no one reads because of its size while some have no plan. Ideally, the plan should be readable; it should be simple and straightforward. Some may argue the plan should include sections on policy, standards, procedures, programs, and practices, which may come into play during a pandemic or crisis. While this is admirable thinking, it clutters the plan, and not all these documents will be used during a pandemic or crisis.

Therefore, it is best to reference important documents that may be needed in carrying out the pandemic or a crisis plan rather than include them in the plan.

Some example documents that could be referenced include regulatory matters that may be associated with the pandemic or crisis, emergency response plans, and appropriate policy, standards, guidelines, procedures, practices, and programs. These documents should be identified by name, along with the location where they may be found when needed.

A pandemic and crisis team members running into the situation room (the room dedicated to managing the pandemic or crisis) with a hundred-page (or more) document under their arm will not work. It is not effective, and it is time-consuming should a team member need to page through the plan for guidance or direction. Indeed, chances are that the team member has not reviewed or read the plan since the last pandemic or spin-off crises, the last team training, or the last pandemic or last crisis team exercise. Thumbing through the plan and at the same time trying to address a pandemic or crisis is problematic. Without knowledge of the plan, guidance, or direction, the team members wonder, "Am I following the plan? What should I be doing?" So, what is the answer? While the plan must be comprehensive and include the identification of reference documents, it should not be longer than thirty to fifty pages. Where documents are thought critical to the plan, they may be attached as appendices or referenced, but they should not be part of the plan. For example: critical phone numbers may be an appendix along with questions to be asked once a pandemic or crisis is identified.

The question should be asked, "Who is in charge during a pandemic or spin-off crisis?" The leader or their designee is in charge, but they should not be responsible to see that the plan is followed. This is important to repeat: the leader is not responsible for carrying out the plan. The pandemic crisis team should have a facilitator to ensure that the crisis plan is followed. Time was spent in writing the plan, testing the plan, and much thought went into the creation of the plan. In the confusion of the pandemic or crisis, the plan may be left on the shelf or be present in the situation room and not followed unless someone puts the plan in front of the pandemic or crisis management team and makes sure that it is used. One would not expect the leader to be thumbing through the pages of the plan to ensure that the team is following the plan. He or she should be involved in decision-making and discussion of the issues so that the best and most informed path forward is taken.

Therefore, the facilitator or his or her designee stands in front of the pandemic or crisis management team, oversees the meeting, and ensures that the plan is being followed. The leader or their designee leads the discussions that result in informed decisions. The plan should call for the identification of a pandemic or crisis team spokesperson, who should be a member of the pandemic or crisis management team, but not the team leader or the facilitator. Since the leader is expected to have all the answers, the media could ask questions that the pandemic or crisis management team has not yet addressed. Therefore, since the leader is expected to have the answers to all questions, he or she can be easily embarrassed by not knowing the answers. Simply stated, when the leader is asked a question and they do not know the answer, the response from the person asking the question could be, "Are you not the leader of this organization? You should know the answer to these questions. Why don't you?" It is for this and other reasons you do not want the leader speaking to the media. A spokesperson for the organization is appropriate.

With one person providing information to the media, the messages should be consistent. When more than one person is providing information to the media, there may be differences in how information is presented that may cause problems of confusion and inconsistency of messaging. This is especially a problem when the presenters are all considered experts. For example: people are saying, "I heard two different things. I do not know who to believe." In addition, there may be a problem if opinions are requested and given. Opinions may vary from expert to expert. This again reflects inconsistency of messaging that the media may characterize as confusing to the public. Those providing the information to the media should avoid providing opinion.

The spokesperson may need to periodically leave the situation room to address the media; however, the facilitator, who is running the meeting, should not leave the room to speak with the media. If the spokesperson is asked a question at the media briefing or at a press conference, they can easily say, "Good question. I shall get back to you with the answer" or "I am certain that very issue is being discussed while I am here with you, and I need to return to that discussion."

When the spokesperson returns to the team meeting, he or she explains what was discussed along with the mood of the people at the briefing, the nonverbal communication that was experienced while speaking with the media, what was learned, and what questions the media asked and they expect answered at the next meeting with the media.

Identify a scribe. This person keeps a time log of pandemic or spin-off crisis team's actions. This will become very important when a critique of team actions is reviewed at a later date when questions like what went right, what went wrong, and what was learned arise. As with any plan that is designed to be streamlined and effective, definitions are critically important. An example is the definition of the word "policy." It is not unusual to have a group trying to create a policy or standards during a pandemic or crisis. In addition, team members may have different definitions of the word "policy." People may say, "We have a policy that addresses that issue," when they mean, "We have a standard or guideline that addresses that issue." The definitions section becomes important when people are calling a situation a pandemic or addressing a spin-off crisis when it does not meet the definition of a pandemic or spin-off crises.

The definition section should also define words and phrases like maximum level of impact, stabilization of the event, remediation, cleanup, recovery, and closure. Actions that may be taken during a pandemic or spin-off crises should also be identified, along with a pandemic and crisis management system. Further, the definitions section should define words like situation room, assessment of damage, and community impact. Within the body of the plan, the roles and responsibilities of team members must be clearly defined.

This becomes important if a member of the team cannot make the meeting and another member of the team needs to satisfy their responsibilities.

The pandemic or crisis management team members should be identified by function and not by name. This would include the team leader and their designee in the absence of the leader. Ad hoc members should be considered. In this way, if the pandemic or crisis requires a specific expertise, they may be included as an ad hoc team member. This may include legal, if not already a team member, IT, medical, engineering, and the like.

The pandemic and crises flowchart, figure 4, should be posted in the situation room for the pandemic or crisis management team to see. In this way, a surfacing pandemic or spin-off crises occurring because of the pandemic cannot result in a loss of pandemic focus. This occurs when the pandemic management team moves to satisfy a spinoff crisis instead of focusing on the pandemic. See figure 1 below.

Pandemic/Spin Off Flow Chart

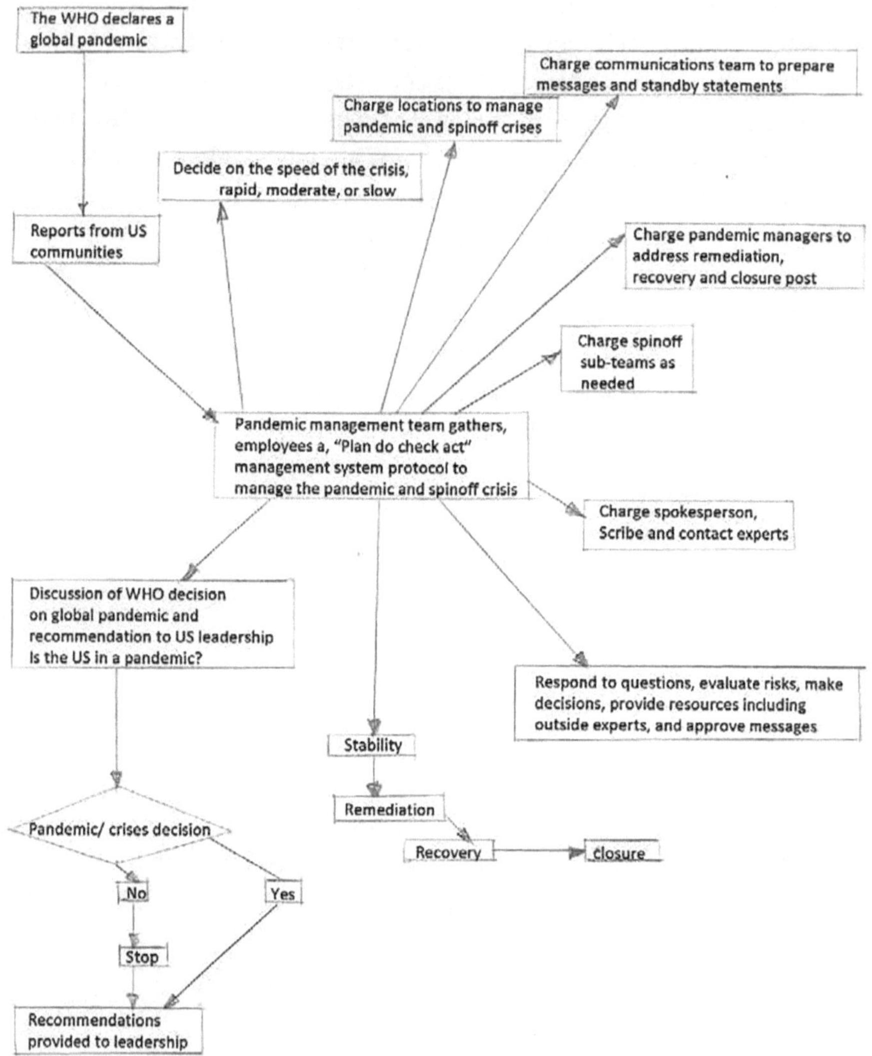

Figure 4

The pandemic flowchart graphic begins with the information on the pandemic coming from the World Health Organization. As crisis spin-off issues begin to surface because of the pandemic, if it is decided that the information being gathered does not represent a pandemic or crisis, then full stop; the arising issues should be handled by others, not the pandemic or crisis management teams.

If a spin-off issue is declared a crisis, the diagram identifies the development of a standby statement and the assigning of the management of the crisis to a team other than the pandemic management team. The leader of the pandemic team assigns the leader and members of a spin-off crisis management team. The specific crisis team leader will report to the pandemic team leader. Further, the chart addresses the speed of a pandemic or crisis and the acquiring of resources. The use of the flowchart permits the team to understand where they are in managing a pandemic or spin-off crises.

It is recommended to use a template to develop or improve your pandemic or spin-off crisis management plan (appendix 4). Using the template, you can fill in the blanks to develop your plan. If a plan already exists, you may use the template to evaluate your plan and, where appropriate, introduce the recommendations of the template to improve your plan. Appendix 4 may also be used to fill in the specific titles and responsibilities that may be appropriate for improving your pandemic or crisis management plans. In addition, appendix 4 may be used to identify how you will use your pandemic or spin-off crisis management document.

Plan content and suggested wording: The following is a typical pandemic or crisis management plan. It is intended as a model for possible use as is or with modifications.

Pandemic management plan

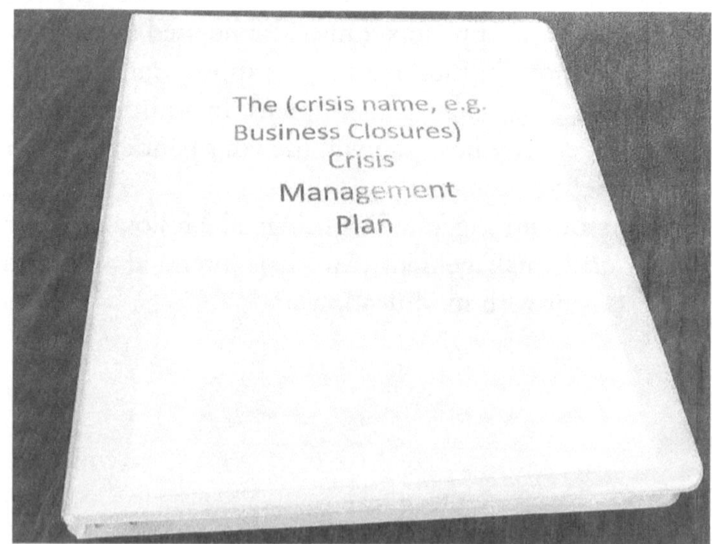

The crisis management plan by name, e.g., business closures

Sample pandemic and crisis management plans

The pandemic management plan is focused on addressing the pandemic and bringing the pandemic to stability. The plan also addresses post-stability issues. Please note that the crises management plan follows the same template as the pandemic management plan. Once the type of crisis is identified, e.g., business closures, the crisis management plan follows the template and includes specifics about the spinoff crisis so it may be addressed and stabilized. The plan also addresses post-stabilization.

Purpose. To enable the clear, effective, and efficient response to a pandemic and all subsequent crisis that spin off from the pandemic.

Scope. The scope of this plan includes a response to the World Health Organization's identification of the existence of a global health pandemic and all spin-off crises that develop because of the pandemic.

A summary of the stages of a pandemic addressed earlier include:

Stage 1: the event, the World Health Organization (WHO) considers declaring a global health pandemic.

Stage 2: the WHO's rapid gathering of information to determine the local seriousness and significance of a pandemic.

Stage 3: the WHO considers identifying a global pandemic and resulting spin-off crises.

Stage 4: a pandemic and possibly a spin-off crisis is declared by the WHO, and the effort begins to minimize the impact of one or both and bring about stability.

Stage 5: stabilization of the event.

Stage 6: assessment of community impact post-stabilization.

Stage 7: remediation and cleanup.

Stage 8: recovery—the returning to normal or as close to normal as possible.

Stage 9: closure—to bring the matter to a close.

Reference documents, including a pandemic and crisis management systems, should be named and their location, identified in appendix 5.

Definitions section

Definitions are critical at times of a pandemic or spin-off crises. When someone is referring to a pandemic or crisis or some other descriptive word, it is important that everyone is thinking of the same definition of that word. The following is a list of terms that may be part of the definitions section of your plan:

Pandemic
Crisis
Pandemic management training
Crisis management training
Pandemic and crisis management flowchart
Communications center
Pandemic and crisis management resource lists
Guidance available to address a pandemic and a crisis
Situation or crisis management room
Assessment of damage and community impact
Maximum level of impact
Life of the pandemic and/or crisis
Remediation and cleanup
Records
Recovery
Closure

The above list may be modified to include additional definitions. Membership: We need to consider memberships in two committees or groups:

1. The pandemic management team membership that, if of value, may be taken in part from the World Health Organization management team.

2. The pandemic spin-off crisis management team member-
    ship—this depends on the type of crisis.

The pandemic core management team reports to the government leadership.

Possible membership identified by title:

Chairperson, someone with a health background, but with
excellent management and leadership skills
Director of external affairs and communications
Director of health emergency program, the plan owner and also
acts as plan facilitator
Government chief scientist
Director of program addressing antimicrobial resistance
Director of emergency response
Director of national health
Director of communicable and noncommunicable diseases
Director of medicines and health products
Director of healthier populations
Possible ad hoc members to the pandemic management team
Director of legal
Director of IT
Director of engineering

Crisis management team membership should include people
with knowledge, ability, and skill sets in the spin-off crisis, such as
job loss, loss of income, depression, no food, purchases limited due
to impact on supply chain, loss of home, cannot pay rent, loss of
car due to nonpayment, cannot pay insurances, loss of medical coverage, and more. The team members should be listed by title and
should begin with the core team. Suggested members of the core
team include the following:

The spin-off crisis management team will be issue focused and
will be charged by the pandemic management team. For example:
crisis identified as lack of materials due to supply chain issues.

Possible membership:

The leader a manger with involvement in trade or transportation
The senior environmental, health, and safety manager, possible owner of the plan
The senior communications manager
The senior legal counsel
The senior person in charge of human resources
The senior person in charge of EPA
The senior OSHA person
The plan facilitator to ensure the plan is followed

Possible ad hoc members, depending upon the crisis situation, include the following:

The senior person in charge of medical
The senior person in charge of distribution
The senior person in charge of transportation
The senior person in charge of insurance
The senior person in charge of information technology
The senior person in charge of engineering

The facilitator, usually environmental, health, and safety background, trains the team in pandemic and crisis management. Other possible core and ad hoc team members may be identified; however, this depends upon your organization.

Example of responsibilities

The following identifies the responsibilities of the members of a typical crisis management team.

Core member responsibilities are similar to those of the pandemic management team. The team should identify a spokesperson and a scribe that will have similar functions to that of the pandemic management team. In some cases, the senior person on the pandemic management team will not be available to serve on a crisis management team. In that case, the senior person serving on the pandemic management team will identify the person who will serve on the crisis management team.

Procedures and other documents. These are existing organization documents that will be used by the pandemic and crisis management teams. They include policies, standards, guidelines, procedures, practices, programs, and the like.

Records. Records refer to the information that becomes part of the event and subsequent crisis situation and how it is managed.

Training. The members of the pandemic and crisis management team need to undergo training. This training will take on many forms—for example, media training, sensitivity training, training in how to work as a team, training with regard to the content of the crisis management plan—and they will need to know what is expected of them. The team backups must be trained so that they cannot only fill in due to the absence of the primary team members but also to relieve the primary team member after twelve hours of activity in addressing the pandemic or spinoff crisis.

Scribe. A scribe is assigned to keep a record, date, time and the like of each action, event, or issue that arises during the pandemic or spin-off crises. This is important when the pandemic and crisis management review takes place post-pandemic. It will help in putting together a timeline from which learnings will take place.

Appendices. These are attached and are referred to within the pandemic and crisis management plan documents.

Document management. This is how all the organization documents will be managed so that those documents that are needed by either the pandemic or crisis management teams be available at times of a pandemic or crisis.

As point of guidance, the following are the titles of people on the World Health Organization leadership team.

1. Chef de Cabinet / Chief of Staff
2. Director for external relations and governance
3. Director, World Health Organization health emergencies program

4. Chief scientist
5. Assistant director-general
6. Assistant director-general, antimicrobial resistance
7. Director-general's envoy for multilateral affairs
8. Assistant director-general, emergency response
9. Special adviser to the director-general
10. Assistant director-general, emergency preparedness and international health regulations
11. Assistant director-General, universal health coverage / communicable and noncommunicable diseases
12. Assistant director-general, access to medicines and health products
13. Assistant director-general, World Health Organization's office at the United Nations in New York
14. Assistant director-general, business operations
15. Assistant director-general, universal health coverage / healthier populations
16. Assistant director-general, specific adviser to the director-general, strategic priorities
17. Special adviser to the director-general
18. Senior adviser to the director-general, organizational change
19. Special strategic advisor to the director-general

CHAPTER 12

The Activities of the Pandemic Management Team during a Pandemic

Many have asked if a team approach is the only way to address a pandemic or spin-off crisis. The answer is no; however, the team approach is the best way since multiple disciplines are needed in addressing a pandemic or spin-off crises, and this can be satisfied with the team approach. The idea of working together as a team for a common cause provides strength, an esprit de corps, determination, and direction. A team that puts in the

T ime and
E ffort can
A ccomplish
M uch

What is a team? The Japanese concept of team is often lost outside of Asia. Other parts of the world tend to call most groups working together a team; however, this is not necessarily what Japan had in mind with the concept of team. In Japan a team is a group of people with different types of expertise invited to come together to solve a problem. The operative word here is "invited." The rank of the individuals, their position, does not matter when it comes to a team. All that matters is that they can work together to solve the problem, the reason that the team was formed. To be asked to be a member of a team is an honor. As with any team, there will be interactions and dynamics to address and personalities that often dictate how a team will interact. While dominant personalities will come to the surface, the team leader needs to be able to ensure that these personalities do not lead the team down the wrong path. The team leader, usually an executive-level person and not the facilitator, should have the skills to prevent any single person from driving a team.

Everyone in the situation room, the location where the pandemic and spin-off crises are being managed and working on the pandemic or crisis, is there because they bring something to the effort, or they should not be there. Each member of the team must have a voice in managing the pandemic or spin-off crisis. To ensure that those who gather function as a team, the pandemic and crisis management teams should undergo team-building exercises, participate in pandemic and crisis management training, and engage in pandemic and crisis exercises.

The following is an example of how a team should work:

A car company wanted to increase the number of cars produced per hour to introduce a new model. Management located the area of an assembly line where the most time was taken to produce an automobile. It was determined that if the company reduced the time to put lug nuts on a car from 120 to 60 seconds, they could produce the new car they wished to bring to market.

They identified team members and a team leader to accomplish the task.

Members of the team included:

- the senior vice president of the car division, the team chairperson
- the vice president of manufacturing
- the director of research and development
- the director of corporate engineering
- the manager of the assembly line
- the supervisor of the assembly line

- the worker who put the lug nuts on the wheels in mounting the tires

It may come as a surprise, but the most important member of the team was the worker who put the lug nuts on the wheels. The reason is that if this person could put the lug nuts on faster while remaining safe, the new model could reach the assembly line. The other team members asked the member who put on the lug nuts to identify anything they could do in the way of making physical changes to the assembly line that would reduce the amount of time it took to put lug nuts on the wheels. Three weeks later, and with the help of the worker, engineering had created a new system for placement of lug nuts and wheels of the vehicles on the assembly line.

The worker reported, "We have shaved sixty seconds off the time, and I am confident, with the couple of additional planned changes, we shall be able to cut another ten seconds off the time." The team was successful, and every member did their part to contribute. The message here is that every member of the pandemic and crisis management teams is selected so that they may bring their expertise to solve problems associated with a pandemic or spinoff crises. While organizations will vary, in general, a pandemic or crisis management team should include core and ad hoc team members. The core team will select and bring on the ad hoc members who can best address the pandemic or crisis. For example: if members of community are experiencing shortness of breath because of the virus, members of the CDC medical staff may be appropriate ad hoc members named to the pandemic management team.

The pandemic or crisis team core members are expected to execute the plan. Specifically, they are expected to assess the potential scope of the pandemic or crises that spin off from the pandemic, help coordinate emergency response action, support, and provide resources, address issues of concern, and act and be perceived as acting responsibly. As with most committees or teams, the more members of the team, the more difficult it is to work together and arrive at decisions. Each team member, core and ad hoc, should have a copy of the pandemic and crisis management plans, and each team member should have a backup who can contribute to the team in the absence of the regular member.

The pandemic or crisis management team activities during a pandemic or crisis

The team members are gathered in the situation room, or they are participating via telephone or video conference. Typically, a pandemic or crisis team member participates by telephone if and only if they cannot make the meeting at the time identified. The team has gathered at the call of the organization communication manager, who is responding to the pandemic WHO declaration. All members should gather within one hour of the pandemic or crisis declaration, and each team member has a specific set of responsibilities.

The pandemic or crisis management team members are encouraged to take the organization's pandemic or identified crisis management plan, e.g., business closures to the situation room if it is some thirty to fifty pages in length to be used as a reference, but not as a crutch. Larger plans should not be brought to the room since too much time will be taken trying to find something in the plan. The plan should not be needed by members of the team since there is a pandemic or crisis management team facilitator who is responsible to know the plan and make sure that it is followed. Pandemic or crisis management team member responsibilities: three-by-five cards identifying the key responsibilities of each team member are kept and brought in by the team facilitator so that member responsibilities may be easily identified. These cards should be provided to the pandemic or crisis management team members by the facilitator as they enter the situation room or via e-mail or fax to members who are on the telephone or participating by teleconference. If core team members are not present, their responsibilities need to be addressed, and the three-by-five cards with their responsibilities of missing team members will be addressed by those members attending the meeting. A listing of responsibilities may be provided as print out instead of three-by-five cards if desired.

Throughout this text, I refer to pandemic and crisis management teams. These are different, and I shall explain. As management of the pandemic moves forward, issues arise that are spin-off issues from the pandemic. The concern is that the focus of the pandemic management team will be forced to change to an issue that is a spin-off from the pandemic. For example: Shutdowns due to the pandemic will affect the economy. People will be out of work, and bills will mount, with no money to pay them. The pandemic management team must not focus on this situation that is indeed an identified crisis related to the pandemic. It is recommended that crises management teams be formed with knowledgeable people to address these crises. Information on what is being done to address the crises should be periodically shared with the pandemic management team. In this way, the pandemic management team stays focused, and the crisis of shutdowns and loss of money will be addressed by other competent people. The crisis management plans very much mirror the pandemic management plans. Therefore, periodic exercises are appropriate for the pandemic and selected spin-off crisis management issues that may arise.

With the pandemic or crisis management team gathered, the team leader, the organization leader, welcomes the members and calls the meeting to order. He or she identifies who is not present or not participating via the telephone or teleconference. Where people are not present or not on the telephone or present via teleconference, the facilitator is asked to provide information regarding the responsibilities of the missing people so they may be shared among those who are present attending via videoconference or on the telephone. However, every effort is made to see that the pandemic or crisis team members or their designated alternate, especially the leader or his or her alternate, can participate. In the absence of the leader or their alternate, the pandemic or crisis management team selects a leader from the members present. Those on the telephone or videoconference should not be selected as the team leader.

With all team responsibilities addressed, the WHO should be contacted for the latest information. The leader of the pandemic or crisis management teams should turn the meeting over to the pandemic or crisis facilitator (assumes a spin-off crisis has not yet been identified) to carry out the plan, and as a first step, the form that was completed during the initial conversation with the WHO is passed out to members or e-mailed to those not in the situation room. The facilitator calls for the identification of a scribe. This person will take notes on all activities and information associated with the pandemic or crisis. It is important that the times and dates of the information and activities be identified. This record and timeline will more than likely be most helpful post-pandemic or crisis.

If it has not already taken place, the facilitator asks the communication leader to have the communications sub-team begin drafting possible standby statement and messages for the media. The pandemic or crisis management team needs to select a spokesperson that is trained to handle the media. They should have good presence; it must be someone that can be trusted by the public and someone who can deliver the messages in a calm, deliberate, and effective manner.

The pandemic and crisis management teams also need to identify managers. The team may want to wait to determine who would be best to serve this role. The selected person should be a leader and a decision maker in a management or leadership role that has knowledge of the pandemic or crisis and can work with sub-teams. The pandemic or crisis managers will oversee pandemic or crisis management activities post-pandemic or crisis stability.

A description of the pandemic or crisis needs to be created and placed in front of the pandemic or crisis management team so they remain focused on the pandemic or crisis and not be pulled in various directions because of new information or crises. One may ask why a written description of the pandemic or crises spin-offs as they occur and why the description is posted in the situation room. A pandemic situation, especially the early stages, is dynamic, and changes will occur as the pandemic unfolds. If not careful, new information arriving at the situation room may result in some of the pandemic management team focusing on the new information and not a pandemic but a crisis spin-off because of the pandemic, hence creating numerous crisis situations—each of which will need resources so that they may be satisfied. These fragmented efforts will dilute the main focus of the pandemic management team. What is needed is the formation of crisis management teams to address each of the spin-off crises that surface. With the written description of the crisis in front of the team members, focus should remain on the pandemic or crisis description and not wander to other issues arising because of the pandemic. This does not mean that the pandemic or the spin-off crises will not change nor does it mean that important new issues arising will be missed by the pandemic or crises management teams. What it means is that as things change and the issues of the pandemic change, every pandemic or crisis management team member will be made aware of the changes, and any modification in the pandemic or crisis description will be shared. The effort in satisfying the crises is and will remain a team effort because of the posting.

Facilitator

As appropriate, modifications to the pandemic or crises descriptions will be provided to the pandemic or crisis team members. The teams should be reviewing the information gathered from the initial phone call with the WHO and try to understand the needs and deliver on those needs. The facilitator should ask the pandemic management team, based on the information to date, what ad hoc members of the pandemic and crisis management teams should be asked to join the teams. This identification of ad hoc members will be dependent on the pandemic or crisis situation and how it develops. Ad hoc members may be added to the team or retired from the team as the management of the pandemic and spin-offs from the pandemic evolve. Using the information gathered from the questions asked of the spokesperson from the WHO when the pandemic was reported, the teams need to work to satisfy the needs of the pandemic and spin-off crises.

For example: Do they need resources? Who, what, when, and where are the resources needed? Where are the resources within the government or outside firms that could be hired to help? How quickly can they get to where they are needed? The teams should also begin reviewing outside resources that could be brought to the event or provide advise via teleconferencing or telephone.

Hopefully, there is a list of these resources, their availability, their cost, and how soon they could be available. In some cases, a list of outside contractors who have been screened and are knowledgeable in the handling of specific pandemic and crisis situations are available. This list should be provided to the members of the pandemic and crisis management teams by the team facilitator.

What is to be done about the media? Are human resources people involved in helping the families of those impacted by the loss of job crisis, or are they needed? Have regulatory agencies been contacted? Many other needs, in addition to those mentioned above, must be addressed in managing a pandemic and spin-off crises. Another question to be answered soon after the pandemic management team gathers is, has the standby statement been issued? If it takes an hour for the pandemic or crises management teams to gather, chances are, a statement has been issued to the media, or the media has written and published something without all the information. The pandemic and crisis management teams need to see the standby statements if or if not issued to the media. If not issued, the pandemic or crisis management teams may make recommendations, but it should not delay the release of the standby statement.

The speed of the crisis becomes a factor in the actions of the pandemic or crisis management teams. Is a crisis developing rapidly? Is the crisis developing at a moderate speed? Is the crisis developing slowly? How are the actions of the pandemic or crisis management teams influenced by the speed of the crisis?

A rapidly developing pandemic may be defined as the following: Rapidly rising number of COVID-19 cases, a rapidly rising number of hospitalization and deaths. This may be caused by COVID-19 or a COVID-19 variant. It must be remembered that the situation with the virus is fluid, not static. Information is continuously flowing from all parts of the world. It is of value to know that 90 percent of all scientists that ever lived in the history of the world are living and working today. Everything needs to be evaluated. In some cases, the information/data has not been peer reviewed. It may get to the media in raw form, and a story is written and published without expert evaluation and input. This is a problem. This is fodder for misinformation. Good scientific protocols must be followed to create and maintain confidence in the process.

A pandemic developing at a moderate speed may be the following: a moderate developing pandemic would be a variant having been identified in another part of the world and is expected to arrive the United States. This is very much unlike the situation from the 1918 pandemic, where people were not flying all over the world spreading disease, but large numbers of troops were traveling in close quarters off to World War I. In 2020 people travelled by air, which caused transmission of COVID-19 worldwide.

Training and a slow-developing pandemic may include the following: a new virus variant has been found in another part of the world, but the possible impact has not yet been determined. Determining the speed of a pandemic will impact how fast resources need to be deployed. If a pandemic or subsequent crisis is developing rapidly, line up the needed resources to travel wherever they are needed. Make WHO and CDC information available to area hospitals, where patients are being admitted. Be sympathetic and empathetic with family members of the dead and severely ill. The pandemic and crisis management teams may need to decide who should manage the pandemic and crises and where it or they should be managed (figure 4). While the pandemic is usually managed at the organization center, where there is a rapidly developing pandemic or crisis, the pandemic or crisis management team will direct activities. Where a pandemic is developing at a moderate or slow speed, it is usually managed by the pandemic management team. The pandemic management team will assign those who will work on the crises. The crises may be managed at the organization level or at locations where it may be best managed, e.g., large numbers of people unemployed in specific areas of the country.

The pandemic and crisis management teams work together to bring the crisis to stability and help supply resources and assign pandemic spin-off crisis sub-teams. The pandemic and crisis management teams may need to function around the clock for a rapid developing pandemic or a few hours as appropriate a day for a moderate or slow-developing pandemic or crisis. The pandemic or crisis management teams should not work more than twelve continuous hours. The alternates to the pandemic or crisis management teams should take over if the pandemic or crises go beyond twelve hours. Each core and ad hoc team member must have an alternate to take over as needed. If the crisis goes beyond twenty-four continuous hours, the original teams return to take over.

In the management of anything, you have four stages: plan, do, check, and act. The check is the measurement of the success of the effort. If you cannot measure what you are doing, do not do it! You may ask, why? Simple. If you cannot measure what you are doing, you will not know if you are making progress nor will you know when you are finished.

Management stages

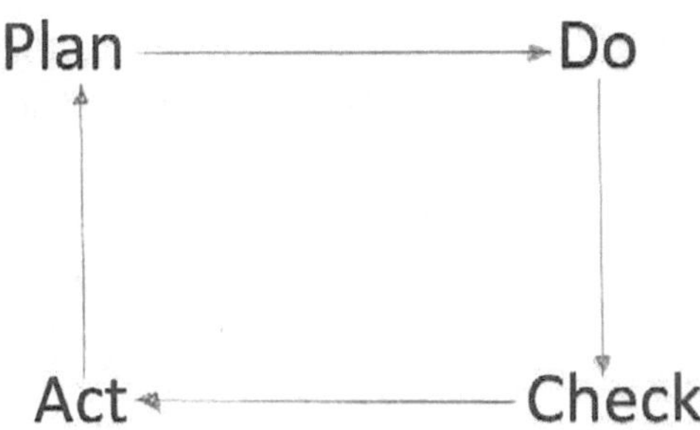

The pandemic or crisis management plans should include a "plan, do, check, act" management system approach. The team should set goals and objectives in working to bring the pandemic crises to stability. Details of using a management system approach in addressing a pandemic or crisis are discussed in chapter 14.

The pandemic and crisis management teams must have a way to receive ongoing updates about the pandemic or crises so that they may respond, not react, to the situation. Reactive behavior is one step from defensive behavior, and when defensive behavior is reached, the organization is in trouble.

Misinformation that disrupts progress in satisfying the pandemic must be addressed immediately, and in no uncertain terms by the pandemic management team or a spin-off crisis management team.

If a conspiracy theory or misinformation surfaces on Monday morning, it must be dealt with by Monday night and dismissed as incorrect or wrong information. If the pandemic management or crisis management team does not respond immediately, the misinformation will not be challenged and, as a result, gain traction.

As a result of the Delta variant impacting the vaccinated and unvaccinated, the CDC has recommended a return to the wearing of masks. This has been met with confusion and criticism. It might be of value to respond to critics with an analogy. If a person builds a porch and attaches it to their home, they use temporary wood supports to hold up the roof until the roof is placed on steel pillars that are in concrete. Then the wood supports are removed. Let us assume the porch concrete is impacted by an earthquake. The concrete cracks, and the roof is in jeopardy of collapsing. You would put back the wood roof supports until the repairs to the concrete can be made, and then the wood supports will be removed.

Basically, the wood supports are similar to the masks that are being brought back to counter the impact of the Delta variant. Keeping in mind that the masks block entry of the virus to the body and the spread of the virus, the criticism does not appear to be justified. Since the CDC did not respond in a strong and effective manner as to why masks and how effective they are in preventing the spread of the virus, they have lost creditability.

With resource needs, the question of who should manage the pandemic, the speed of the pandemic, and the standby statement addressed, the pandemic management team or the crisis management teams need to focus attention on the development of messages. Such messages consist of one sentence where the organization wants to make a clear and strong point that it wants remembered. There should not be more than three to five messages provided the media, preferably three at any point in time. The technique of delivering the messages involves answering a reporter's or interviewer's question and then introducing the message you want heard. The message is introduced without taking a breath. The spokesperson should look for opportunities to introduce the messages several times, whenever and wherever possible. In developing the messages, it is important to identify first the audiences and stakeholders for the messages.

Considering this is a global pandemic, this could include government employees, the community, the country, or the world. When developing the messages, it is important to state only facts and not offer or state an opinion. Show the government's approach in a good light. For example, and if possible, identify the accomplishments of the organization, CDC, and the like, and speak in sympathetic and empathetic terms, especially to those families of those who have died due to the pandemic. Demonstrate a caring organization totally committed to addressing the pandemic and spin-off crises.

Television interviews should be conducted with the organization's spokesperson, and radio sound bites should also be from the spokesperson. Positive unsolicited testimony on television or radio or other forms of media about the quality and sincerity of government's approach to addressing the pandemic will be viewed as very positive. A second person, in addition to the spokesperson, should be present at any media interview. This person should not speak on behalf of the organization. They should act as an observer and share their observations and information with the spokesperson and other senior management post-interviews. You may wonder, why a spokesperson and not an expert in a leadership position? Leadership is expected to have all the answers while the spokesperson is not expected to have all the answers. The spokesperson can say, "Good question. I do not know the answer but will get the answer and provide it to you at the next briefing." Leadership cannot get away with such a response. The leadership not knowing the answer implies lack of knowledge and leadership.

There is a strategy for delivering messages. The messages could be delivered as part of a brief opening statement from the spokesperson at a news conference. The technique to be employed by anyone responding to the media is to first answer the question being asked with fact, not opinion. Following the answer to the interviewer's question and without taking a breath, the spokesperson should present one of the three messages. If possible, deliver each of the three messages after each of the interviewer's questions are answered.

One may ask, why only three messages? The answer is that there is not enough time to deliver more than one sentence after responding to an interviewer's question. More than three messages will be confusing, and people will not remember the messages. In addition, if there are too many messages, people will not know which is the most important. You want the audience to remember what you are saying; therefore, limit what you are saying to key points that the audience will remember. The following are examples of three messages that may be appropriate depending on the pandemic or crisis and the situation:

1. "We are deeply saddened at the impact of the pandemic. We are working with the families who have lost loved ones and are doing everything humanly possible to ease their pain."
2. "Our organization has had an exemplary performance in addressing pandemics for example aids."
3. "We have brought together all the experts that we could gather from academia, industry, and government to work with us to bring this pandemic and spin-off crises to a successful and rapid conclusion."

The messages, which must be approved by the pandemic and crises management teams, should focus on five areas. These include positive attributes, functional positive consequences, feelings, emotional consequences, and values at the highest level.

Before the pandemic or crisis management teams can achieve stability, they must define stability for the pandemic and crises. Stability may differ from pandemic to pandemic and crisis to crisis; however, when stability is reached, it must be recognized and agreed to by all members of the pandemic or crisis management teams. With stability in place, the crisis management team may begin the remediation effort, if not already underway, and begin to review the effort to bring about recovery. The pandemic management team may charge the sub-teams not already working. The pandemic management team and possibility the crisis management team may fall back and permit the sub-teams to work under the direction of the pandemic manger and the crisis managers. The pandemic management team will reconvene at the request of the pandemic or crisis team managers or organization leader.

CHAPTER 13

The Fundamentals of Managing a Pandemic

The guiding principles and fundamentals of pandemic and crisis management is simple, first and foremost—save lives. One way of determining whether a pandemic or spin-off crises exists in your country is to use a matrix. The matrices, figures 5 and 6, are usually made up of boxes. The X axis addresses the severity of a possible pandemic or spin-off crises, and the Y axis addresses the probability of the event that has occurred, representing a pandemic or spin-off crises. The Y axis identifies the probability of occurrence as a percentage, 20, 40, 60, 80, and 100 percent. The severity of impact, the X axis, may be described as very low, low, moderate, high, and very high. Terms such as "likelihood" and "consequence" may be substituted for probability and severity if so desired. The matrix may be presented where either the bottom right or the top right corner identifies the worst case. In the matrices to follow, figure 5 is presented with the bottom right as being the worst case, and figure 6, the top right corner is the worst case.

The second matrix, figure 6, is a five-by-eight matrix. The boxes left to right on the X axis indicate increases in the level of severity of the pandemic or spin-off crises. The boxes bottom to top on the Y axis indicate levels of the involvement in the pandemic or spin-off crises, e.g., the local community hotspots, the state, or the country.

The more levels of involvement in the pandemic or crisis, the greater the possible impact on the general public and the government. The numbers on the Y axis indicate the number of levels of involvement. For example, and using the levels of involvement described above, there would be three different levels. While figures 5 and 6 to follow are presented in white and shades of gray, it is appropriate to color the boxes of both matrices to indicate increases in severity of impact and probability of occurrence. The dark gray in the figures to follow, in both matrices, indicate the need for a strong recommendation to the organization leader that a pandemic or a spin-off crisis exists. The moderate gray in the figures to follow indicate a need to recommend that a pandemic or a crisis exists. The light gray in the figures to follow mean information should be provided to the leader without a recommendation that a pandemic or crisis exists. The leader of the organization with the input of the pandemic and crisis management teams will make the decision concerning declaring a spin-off crisis exists once all the information is presented and discussed. The white boxes indicate that it is not necessary to meet with the leader of the organization; however, a phone call identifying the pandemic or spin-off crisis is in order, and an ongoing monitoring of the situation is necessary to ensure that the condition does not deteriorate into darker gray where a pandemic or spin-off crisis recommendation would be appropriate. The matrix is designed to be a guide to the group providing the crisis recommendation to the leader of the organization. If it is unclear to the team reviewing the event as to whether or not to recommend that a pandemic or spin-off crisis be declared, the team is to err on the side of caution. The crisis matrices follow.

Decision matrix regarding the United States declaring a pandemic

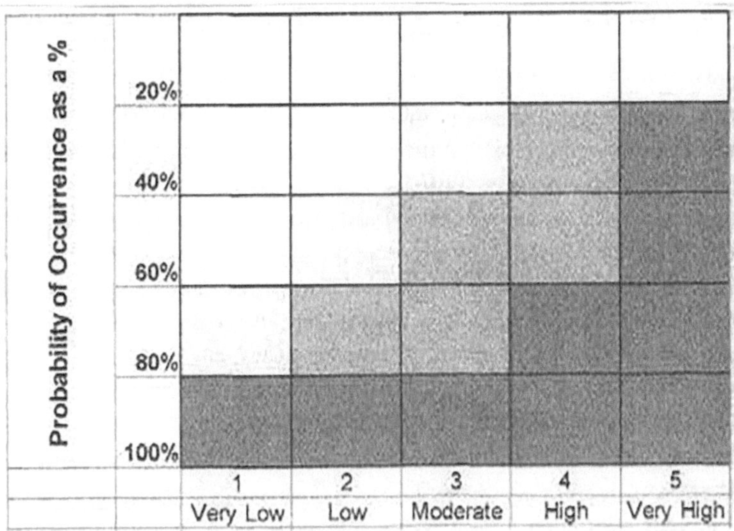

Severity
Declaring a pandemic
Figure 5

The WHO may declare a worldwide pandemic that is based on earlier identified criteria. The criteria used by the WHO can result in a pandemic declaration of a worldwide pandemic without the involvement of North, Central, or South America. The question remains: does the United States declare a pandemic with no or a single case of a virus? The above matrix helps in that decision. The five-by-five matrix includes white, very light gray, gray, and dark gray. Moving from white to dark gray indicates moving toward a US declaration of a pandemic. Across the bottom of the matrix are the words "very low," "low," "moderate," "high," and "very high." On the left side of the matrix is the probability of occurrence expressed as a percent.

Please keep in mind what follows is based on subjective judgements. A United States pandemic may be identified regardless of the use of the following matrix. The use of the matrix will give creditability to the decision regarding a pandemic declaration. Only the WHO can declare a global pandemic.

Figures 5 and 6 may be used to determine if a pandemic should be declared in the United States where there are little to no cases and when the WHO has declared a global pandemic, and the United States is not listed by the WHO as an area where the criteria causing the identification of a pandemic has not been reached.

For this evaluation, I have chosen three areas to be monitored:

- Total caseload attributed to the virus in the United States
- Total hospitalizations attributed to the virus in the United States
- Total deaths attributed to the virus in the United States

If the data is collected and evaluated quarterly, and if I identify one hundred thousand cases in a quarter as the trigger to identify a pandemic exists in the United States, then the probability of occurrence in figure 5 is 100 percent. Therefore, if sixty thousand cases occur in the quarter, we would move up the matrix to 60 percent. Severity will depend on speed of developing cases. If cases are developing rapidly, a subjective judgement of very low, low, moderate, high, or very high needs to be made. Critical here is the box on the matrix we select. We are at 60 percent, and if we move over to very high, then the selected box is in the dark-gray area. Worst-case scenario, a US pandemic should be declared.

We can use the same approach for hospitalizations and deaths from the virus. If we choose fifty thousand hospitalizations in a quarter and we have forty thousand, we are at the 80 percent probability level; and if hospitals are being overrun with admissions and the severity is high, the selected box on the matrix is in the dark-gray area, and a pandemic needs to be declared. The same is the case with deaths reported in a quarter. If we choose perhaps ten thousand deaths in a quarter, and eight thousand occur again, we are at the 80 percent level and very likely in the high severity, and this would also result in a need to declare a pandemic. While I have presented three areas to monitor, there are more if you wish to include them. An obvious question is, if you monitored three or perhaps five areas, and only one of the four was identified in the dark-gray area, should a pandemic be identified? The answer is yes. You should not consider averaging items being monitor when it comes to people's health.

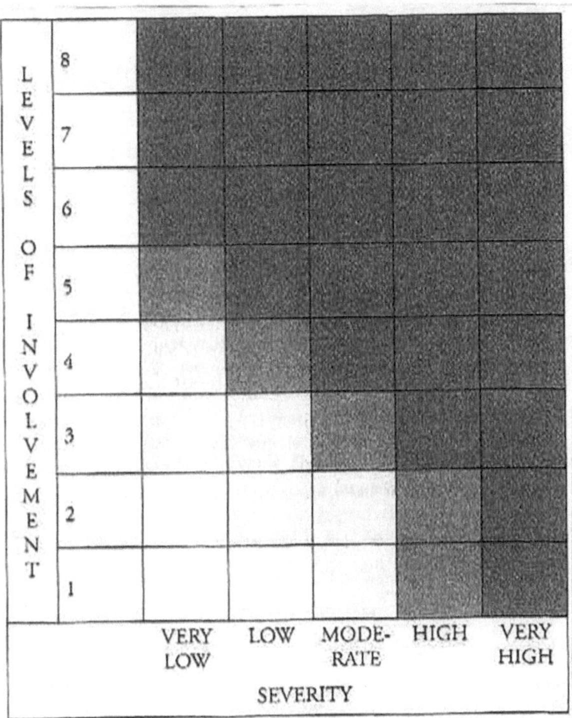

Multiple spin-off crises
Figure 6

The list of possible spin-off crises below are used in figure 6. The numbers on the left side of this matrix address areas of possible spin-off crises. These include in no special order:

1. Business closings
2. People laid off
3. Loss of home or car or both
4. Loss of income
5. School closings
6. Lack of food
7. Cannot pay rent

8. Loss of medical insurance
9. Other

Using figure 6 and assuming the results from figure 5 have not reached the pandemic stage, if any of the list of spin-off issues listed above, for example, business closures, reach a level 25 percent above an acceptable norm, it should be placed on the graphic at position 1 from figure 6 and in the very high dark-gray area. This is a call to declare a US pandemic. It implies under reporting of cases. Should the level number reported for the spin-off issue be some 20 percent above the norm, it would be placed on figure 6 at place 1 and in the high gray area. It has not reached the level to declare a pandemic. However, if a second and third spin-off issue, loss of jobs and mortgage foreclosures, arise, both reaching 20 percent above the norm, both are placed on figure 6 positions 2 and 3 as high. You can see from figure 6 that with three spin-off issues identified as 20 percent above their norm and placed in the high area, the dark-gray area has been reached, and a pandemic should be declared. This is a clear indication that the cases are being under reported. The pandemic management team is gathered and identifies the formation of three crisis management teams to address the spin-off issues: business closures, job loss, and mortgage foreclosures.

Using the two matrices, a strong recommendation should be made to the organization leader and the pandemic management team that a pandemic be declared and the pandemic management team gathered. The recommendation is because you cannot average the impact. To support this thinking, if someone were to drink from five different glasses of liquid, and one of the five contained cyanide, there would not be an average impact. If the organization leader agrees that a pandemic should be declared, the communications representative and their staff are to call the pandemic team together at a pre-agreed location and time. The most senior person with input from the pandemic management team decides whether there is a pandemic in the United States and that there are spin-off crises that need to be addressed. If it is decided that the situation does not merit identifying a pandemic or spin-off crises, all activities and actions associated with the pandemic or spin-off crises are stopped. You may ask, why use matrices? They are tools that may be used if underreporting of illnesses caused by COVID-19 is suspected.

Summary

A global pandemic is defined by the WHO. Each country responds to the WHO identified pandemic for their respective country. Each country will identify the spin-off crises and respond to them. It is best to have criteria, such as the matrices identified above that, once met, a pandemic and possible spin-off crisis situations identified.

Be prepared. First and foremost a government must be prepared to address a pandemic or spin-off crisis situations.

Have a plan. It is critical to have a plan to address a pandemic and spin-off crises. In addition, there needs to be a means to ensure that the pandemic and crises plans are tested and kept up to date.

Have core pandemic and crisis management teams with alternates for each team member. The ad hoc members become part of the team as needed and should also have identified alternates.

Conduct exercises to test the pandemic and crisis management plans. Both plans should be tested annually. This can be paper exercises or more elaborate exercises, which may involve the participation of local emergency ambulance services and police departments.

The pandemic or crisis plan owner. This ensures that the plan will not be changed without the owner's knowledge and the approval of leadership. As a controlled document, changes in the documents must follow a protocol that includes a review and sign-off by each member of the pandemic and crisis management core teams. The plan owner is also responsible for all plan updates.

Have a crisis facilitator. The pandemic and crisis facilitators conduct the crisis management team meetings. He or she ensures that the plan is followed and thereby allows the leader and other team members to concentrate on the pandemic or spin-off crises.

Both plans need to be controlled documents. As a controlled document, the plan will have a known distribution, and copies of the plan will not be shared with people who do not need to have a copy. Further, limiting distribution should also keep the possibility of the plan reaching the media at the time of a pandemic or crisis to a minimum.

Have crisis spin-off teams. These are groups that compile information for the pandemic and crisis management teams. In addition, they do the legwork of contacting people and ensuring the effort to manage is functioning. They also continue to work post-stability into the remediation, recovery, and closure stages.

Have pandemic and crises managers. Once the pandemic reaches stability and the pandemic management team chooses to step back, a pandemic manager is identified. The spin-off crisis manager takes over the crisis when the crisis situation becomes stable. He or she is charged by the spin-off crisis management team, and the crisis management team steps back. The pandemic or crisis management team may be brought back to the table by the managers due to changing situations where the pandemic or crisis team is needed.

Prepare a standby statement. This is the statement that is given to the media so that they have something to report before the details of the pandemic or crisis are available. The intent of the statement gives the media something to write about without their need to be involved in the pandemic or crisis. However, it may not keep the media from involvement.

Have a scribe. It is important to document all information and actions concerning the pandemic and crises. This will help in the critique of the actions in response to the pandemic or crisis, which may take place weeks after the pandemic and crises has reached stability; and from a liability standpoint, it is extremely valuable to have a written history of actions and events responding to the pandemic and spinoff crises.

Understand that a pandemic and a crisis have stages. There are nine stages. Knowing which stage the crisis has reached is critically important to making correct decisions.

A small group of people (three to four) after careful consideration, criteria, and documentation may make a recommendation to the leader, calling for a decision that a pandemic be declared. Only the most senior person in an organization can declare a pandemic. A country health pandemic maybe identified as an epidemic. Only the WHO can declare a global pandemic.

If more than one person can declare a pandemic, there will be confusion and problems. An epidemic, pandemic, and a crisis cost money and takes resources. The leader or their designee is the best person to commit resources and declare a pandemic, epidemic, or crises.

Develop messages. Messages are important as the pandemic and crises unfold. The pandemic and crisis management teams should work together to ensure that the messages provide the information they want to give to various audiences. The messages should be crafted for targeted audiences, e.g., vaccinated and unvaccinated people. These may include families, government, those specifically impacted by the pandemic or a spin-off crisis and the general public, as examples. The messages should be limited to three to five in number.

Identify a spokesperson. The government or organization spokesperson should deliver the messages. He or she needs to be trained in working with the media. During an interview concerning the pandemic or spin-off crisis, he or she delivers only facts, no opinion. In addition, and where possible, they are to deliver the messages approved by the pandemic and/or the crisis management team. A second government or organization person should be present during media interviews to listen, not to speak.

Train people to speak to the media. Only people who have received training on how to speak with the media should speak with the media.

Be in control. Those charged with addressing the pandemic and its sub-crises must be in control. They must present the confidence that the matter is in hand and the appropriate resources are in place and addressing the situation.

Understand and correct behaviors that may lead to the pandemic, epidemic, or crises. Some behaviors may lead to a pandemic spin-off crisis situation. These behaviors need to be understood, and training on how to prevent such behaviors must be offered to all.

Understand the behaviors needed during a crisis. Staff needs to be trained in the behaviors needed to best manage a pandemic or spin-off crisis situations. Being calm and in control is the key to success during a pandemic or crisis situation. Leadership operating in a panic mode will not be effective; they will be detrimental to satisfying the situation. Such behaviors are more likely to make mistakes if they are under pressure.

Maintain an up-to-date resource list. People with skills and talents will need to be called at times of a pandemic or spin-off crises. Many of these people are experts in their field, and having the correct phone numbers or e-mails may be critical to the situation. Such contact information must remain current.

Prepare a list of questions to ask the area undergoing the pandemic, epidemic, or spin-off crisis. See appendix 1. Too often the location experiencing the pandemic or crises is not able to provide the information needed for the pandemic or crisis management teams to provide assistance. A list of questions to ask of the location that will in turn be shared across government that will help in preventing repeat phone calls from many people asking the same questions.

Stay focused during the pandemic, epidemic, or crisis situation. Often during a pandemic, epidemic, or crisis situation, information flowing into the teams will take the teams in a direction not consistent with satisfying the pandemic or crisis. While this information may be important, the teams need to remain focused in attempting to stabilize the situation.

Identify a backup location where the pandemic or crisis teams can meet. The pandemic, epidemic, and crisis management teams should have a place to meet. In addition, there should be a backup location if the first location is not available. Both locations should have specific equipment that will help manage the pandemic or spinoff crises.

CHAPTER 14

The Use of a Crisis Management System

The crisis management system: plan, do, check, and act

There's a difference between management and leadership. As an example, a company was cutting through vegetation in a jungle, and the manager had the effort well in hand. Groups were on twelve-hour shifts; there was a group in the field cutting the vegetation, a group sharpening knives, a group of cutters, and a group of knife sharpeners sleeping. At the end of twelve hours, those sharpening and those in the jungle returned to rest, eat, and sleep; and the group that were sleeping went out to cut the jungle or to sharpen knives. With the system put in place by the manager working well, the leader came to visit. After exchanging a few words with the manager, the leader climbed to the top of the tallest tree in the jungle to see where progress was being made in cutting down the jungle. Upon looking out onto the field, the leader was horrified and yelled down to the manager, "Stop. You are cutting the wrong jungle." The manager looked up at the leader and said, "Shut up. We are making progress." It is clear to see with this example that there is a distinct difference between leadership and management. This is not to say a manager cannot be a leader or a leader cannot be a manager; it simply demonstrates that there is a difference. It is best to define terms before moving forward. The following is a list of terms and their definitions to ensure the establishment of a high-level management system that includes a pandemic and crisis management system.

The hierarchy of management:

- Belief
- Policy
- Standards
- Guidelines
- Procedures
- Practices
- Programs
- Criteria
- Goals
- Objectives
- Other

Belief. This is where it begins. It is an organization's thoughts concerning what they believe about a given topic (e.g., pandemic and crisis management).

Policy. This is the highest-ranking formal document in an organization concerning a specific topic (e.g., an environmental, health, and safety policy, a human resources policy, among others).

Standard. This is a means of carrying out the policy, and it is considered an organization rule. As a point of reference, government legislation may be considered policy, and government regulations may be considered standards.

Guideline. This is a written means to achieve a standard. It may not need to be followed so long as the standard is achieved.

Procedure. This is the "how to" document.

Practice. This is a means to an end and typically has a long life within an organization.

Program. This is a focused effort with a finite life that is carried out within an organization.

Criteria. This is a value used to determine acceptance (e.g., does it or does it not meet the criteria?).

Others. Depending on the organization, there may be many other terms that are used to manage an organization.

This becomes important if someone is speaking about a standard, which is an organization requirement, and the person listening is thinking about a guideline that may not need to be followed. There could be a serious problem, especially if the topic is a pandemic or crisis management. Definitions are important especially in creating policy. For example: an organization manager may say, "We believe that everyone who works here should be safe." A policy statement would be developed from that belief. The policy statement could read, "It is the policy of the [the name of the organization would be placed here] organization to take foremost account of the safety of our employees." Note the policy does not say how the organization will take foremost account, only that they will take foremost account; therefore, the how-to is not part of the policy. A policy, which should be one sentence in length, should be followed by the words, "In carrying out this policy, the organization shall…" Obviously, the words chosen here are important. In this case, the word "shall" indicates a commitment. However, words such as, "plans to" or "intends to" reflect less of a commitment. This statement is followed by a series of bulleted action items explaining how the policy will be carried out. While the entire text is referred to as policy, only the first sentence is the actual organization policy. In the following hierarchy of written organization management protocol, standards are prepared to help carryout policy, guidelines are written to help in meeting standards, and procedures, practices, and programs are written to help achieve the policy. This mindset should be followed from the leader of the organization all the way to the job responsibilities of the person sweeping the shop floor. In this way, everyone can see that their involvement is important and that they are all contributing to the success of the organization. This is especially true during a time of a pandemic or spin-off crisis.

An organization should have a smaller number of policies than standards because policies are the most important organization documents. For example: it is not necessary to have an organization crisis management policy if the organization's environmental, health, and safety policy is broad based, as it should be. It can more than likely include crisis management. If the policy states, "Take foremost account of health and safety and protect the environment," it is implied that every effort will be made to prevent a pandemic or spin-off crisis from occurring. Therefore, management standards, guidelines, procedures, and practices for a pandemic or crisis management must be consistent with the organization's environmental, health, and safety policy.

Policies provide direction, and there should be more standards than policies and more guidelines than standards. Picture a tree as the format. The beliefs are at the top of the tree. The trunk of the tree represents policies. The large branches off the trunk are the standards. Guidelines come off the large branches as smaller branches, and off the smaller branches are even smaller branches that represent the procedures, practices, and programs. This is another way of saying that there will be many more procedures, practices, and programs, then guidelines, standards, and policies. The obvious question here is, why is this important when it comes to pandemic, epidemic, and crisis management? It is important because during a pandemic, epidemic, or crisis, there will be a need to look to an organization's policies, standards, guidelines, and the like as the pandemic, epidemic, or crisis develops and is managed.

Working through an event that has been declared a pandemic, epidemic, or crisis and using a management system approach makes sense. Consider the following example. Several communities across the country have reported a large cluster of COVID-19 cases. The pandemic management team has been gathered, and they are briefed as to why an organization pandemic, epidemic, or crisis should be declared. This is important in that only the leader of the organization with guidance from the pandemic management team can declare a pandemic for the United States. As part of working through a decision regarding declaring a pandemic, epidemic, or crisis, the pandemic management team may choose to establish goals and objectives. Here again definitions are critical. Goals and objectives may be defined as follows.

Goals. A goal is something to plan on achieving as you move the process forward. A goal does not need to be achieved. While you plan on achieving a goal, the idea is to keep a focused direction. For example: the goal of the pandemic management team is to satisfy the pandemic as quickly as possible.

Objectives. An objective must be measurable and attainable. It is very important to achieve objectives. Each objective that is achieved moves you closer to the goal. In other words, small achievable and measurable steps will bring you to your goal.

An obvious question is, "Do I need a pandemic, epidemic, or crisis management system or, for that matter, any management system?" The answer is simple—no. With that being the case, "Why should I consider a management system to address a pandemic, epidemic, or crisis?" The answer is again simple—the management system provides a sense of confidence and understanding since it is an organized approach to achieve the goal. A management system is a structured approach for the management of issues such as a pandemic, epidemic, or crisis. Management systems are built upon feedback loops. These loops allow for changes and corrections should the plan not immediately achieve its objectives and goals. Management systems can be confusing. For most of those considering employing a management system, the question is raised, "Since I already have a system in place that works, why do I need a more formal management system?" A management system permits better focus, and it allows the understanding of the stages of the system and where you are in the system. For example: are you in the plan stage, the do stage, the check stage, or the act stage? This becomes critical as difficulties arise in managing an effort such as a pandemic, epidemic, or crisis. Using "plan, do, check, and act," you can move from stage to stage and use feedback loops for making changes to ensure success.

Your organization may already have a management system in place. You may not recognize it as a management system; however, if you did not have a management system in place, you would have chaos. Assuming you do not have chaos, you have a management system in place; however, what you have is probably not a formal management system, and therefore, it can be improved. The four phases of a management system—plan, do, check, and act—are used to help the crisis manage team manage the nine stages of a crisis.

Feedback loops are critical to the success of a management system. By using scenarios and exercises, feedback loops may be employed to make corrections and updates for the pandemic, epidemic, and crisis management plans. As each stage of the plan is implemented, there needs to be a feedback loop concerning its success or lack of success. Where implementation takes place and the plan is judged to have not been successful, the plan is modified, and a second attempt is made. The feedback loops are used until success has been achieved. The following is an example of a feedback loop.

A standby statement as part of crisis exercise scenario identifying a pandemic went out to the media thirty minutes before a new bulletin identified a spin-off crisis because of the pandemic. There is no mention of this latest event in the standby statement. In addition, the standby statement does not demonstrate any sympathy or empathy for any loss of lives or property. The communications group along with the pandemic and spin-off crisis management teams must focus on the follow-up statement concerning the latest event as soon as possible. The statement needs to be factual and confirm all that is known about the spin-off crisis. This statement must reflect empathy and sympathy for the families of those who died because of the virus. It is the use of feedback loops in scenario development and exercises that permits corrections to the plan to prepare for situations such as the one described above and to be ready should a similar situation occur during an actual pandemic or spin-off crisis.

A second example of a feedback loop: the pandemic management team has assigned responsibility to manage a spinoff crisis. It is best to have someone knowledgeable in areas such as people losing jobs and, as a result, their homes due to job loss because of the pandemic. However, there are issues. For example: the response on the part of the selected leader may be slow. He or she may be difficult to reach. Planned updates concerning the spin-off crisis are not being provided. The second-in-command is contacted and explains that the leader is not at the meeting. He or she took their son to the doctor this morning and is unaware of the status being requested.

The team has tried to reach them, but their cell phone is probably turned off as requested by most physicians' offices so that it does not interfere with office monitoring equipment. The team is certain they will reach him within the hour, but it is unlikely that he will be able to get to the meeting anytime soon.

While understanding that the selected leader is not at the scene and his or her alternate appears inexperienced, the crisis management team using management system feedback loops must take back responsibility for managing the spin-off crisis or find someone else that they believe can do the job. The pandemic or crisis management team returns to the stage of the plan that calls for deciding where the pandemic or crisis is to be managed and makes the change.

A third example of a feedback loop: a toxic material was found in the building housing the organization headquarters. Government officials are on scene and require the evacuation of the building, and it is estimated that it will take three weeks before staff can be permitted back into the building. The feedback loop causes the pandemic or crisis management team to identify a backup location to manage the pandemic or crisis. In an exercise, a crisis management team quickly realized that it did not have a backup off-site location for the pandemic or spin-off crisis management teams to meet. The situation was quickly satisfied after being addressed in the critique of the exercise, and the management plans were modified to include a backup location for the pandemic and spin-off crisis management teams to meet should the primary meeting location not be available. This is only one example of the value of both pandemic and crisis exercises. It is important to note that in identifying this fallback location for the crisis management team to gather, it was quickly determined that the backup location needs to be equipped with much the same materials (e.g., computers, telephones, and the like) as the primary meeting place of the crisis management team.

The next step in the process is to activate or create a crisis management plan using the "plan, do, check, and act" approach of a crisis management system. Appendix 5 explains how this can be accomplished. Hopefully, pandemic or spin-off crisis management plans exist in your organization and have been tested; however, if this is not the case, a plan needs to be created and tested. The "plan" (using a management system) is designed to satisfy the goal and objectives of the organization; however, there can be more than one objective to satisfy the goal. The plan mandate should be the following.

First save lives. This is followed by support of those experiencing the pandemic, epidemic, or a crisis spin-off from the pandemic. Testing of the plan represents the "check" in the plan, do, check, and act phases of the management system. Testing of the plan can be in the form of a paper exercise or an exercise involving outside resources (e.g., local fire and police departments may also be of value).

The pandemic and crisis management plans should include all the elements that are expected in addressing a pandemic or crisis.

The crisis management "plan, do, check, and act" approach should be used as a means of addressing each pandemic or spin-off crisis situation, whether it be real or an exercise to test the plan. A feedback loop enables a review and permits possible changes to improve the plan after each exercise or each time the plan is used in an actual pandemic or spin-off crisis. With the plan changes made and the pandemic or crisis management team satisfied that it is an effective tool to manage a pandemic or crisis, it is recommended that the organization leadership group act—the fourth and last part of the "plan, do, check, and act" management system—and make the updated pandemic, epidemic, or crisis management plans part of the organization list of documents from which day-to-day management takes place. The use of the management system to address a pandemic, epidemic, or crisis situation ensures that the most effective way to manage the pandemic or crisis will be utilized. It ensures that changes to improve the plan will not be missed and that ongoing updated pandemic, epidemic, or crisis management plans will be used to satisfy a pandemic, epidemic, or crisis situation. The following are some examples of management systems that are presently being used. They include ISO 9000, ISO 14000, ISO 18000, and the chemical industry's Responsible Care® program, among others.

CHAPTER 15

Pandemic Stability, Remediation, Recovery, and Closure

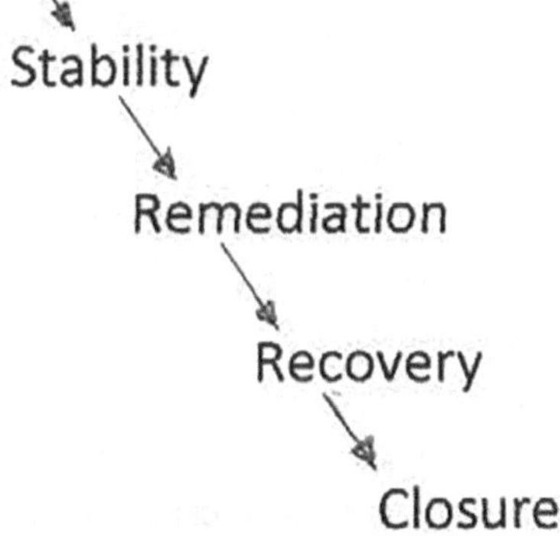

Stability

At some point, a pandemic or spin-off crisis will reach stability. Stability may be defined as a firmness of position or continuance without change, a sense of permanence. Synonyms that are associated with stability include steadiness, strength, soundness, poise, solidity, and balance. Examples of stability as they relate to pandemic and crisis management include the following: the number of new virus infection have slowed to a point of normal number of infections, and they are not expected to increase; the number of new filings for unemployment due to shutdowns and loss of jobs are down; the number of hospitalizations and deaths are down. The pandemic management team must agree when stability is reached after which they may charge a pandemic and crisis manager and a series of sub-teams to continue the work effort if they have not already done so. In some cases, sub-teams such as communications have already been charged to help create messages. While they step back, this does not mean that the work of the pandemic or the crisis management teams is over. The team falls to the background and maintains contact with the sub-team efforts that are underway.

At stability, the role of the crisis facilitator is diminished, and he or she should check to determine if all the elements of the plan have been identified and implemented. When time permits, the facilitator begins to identify what was learned that could strengthen the pandemic, epidemic, and crisis management plans. A pandemic management plan may be used to satisfy an epidemic plan. At stability, the role of the pandemic manager gains in importance, as he or she is responsible for organizing the crisis spin-off or sub-teams and works closely with all of them. He or she acts as the liaison between the pandemic and crisis management team and the crisis management sub-teams and may call back the pandemic and crisis management teams to the situation room at any point judged appropriate. He or she may with the agreement of the pandemic team leader sunset spin-off crisis management teams when their work is done.

At stability, the pandemic spokesperson announces to the media that stability has been reached and expresses confidence that the worst is over. However, pandemics are known to come in waves. It would not be impossible to see another wave of infections of COVID-19 after stability has been declared. This is one of the reasons the sub-teams continue their work effort so that a rapid response to a new wave can be carried out if needed. It is appropriate for the spokesperson to credit the many people who have worked around the clock to save lives and bring the pandemic to a point of stability. For clarity, if a crisis sub-team has a communications person, they work closely with the pandemic team spokesperson. This is to avoid the possibility of miscommunication and communications in a vacuum. The pandemic spokesperson may explain that there is much work yet to be done in the coming days and weeks and commits the organization to see the process through remediation and recovery to closure. Further, the pandemic team spokesperson is the focal point of all communications if the crisis sub-teams do not have a spokesperson.

Remediation

With stability agreed, remediation is the next step in the process. Remediation may be defined as the correction of something bad or defective. It is a reversal or stopping of something that is considered to have impacted the community, the country, or the world. It is important to stop politicizing the pandemic and spin-off crises and stop creating unnecessary fear.

The following addresses the roles of the pandemic facilitator, the crisis manager, the crisis spokesperson, and the crisis spin-off or sub-teams during stability, remediation, and recovery. The crisis sub-teams may include job loss, supply chain issues, need for hospitals and hospital beds, human resources, information technology, a sub-team for business shutdown and loss, legal, and communications, among others.

Pandemic facilitator

The role of the pandemic facilitator in the remediation phase is to work closely with the pandemic manager and ensures that the remediation section of the pandemic and crisis management plans is followed. The pandemic facilitator now attends meetings of the sub-teams and listens as a resource to each sub-team. At recovery, the role of the pandemic facilitator is to ensure that all learnings from the pandemic and spin-off crises have been captured and made part of an updated pandemic and crisis management plan. The facilitator conducts training sessions for all members of the pandemic management team, explaining the changes in the plan because of managing the pandemic or the pandemic and crisis exercises and distributes the updated plans as a control document to the pandemic and crisis management teams.

Pandemic manager

The role of the pandemic manager in the remediation phase of the pandemic or crisis management is to attend each of the crisis sub-team meetings to ensure consistency, nonduplication of effort, and progress. While the pandemic manager demonstrates leadership, he or she does not chair the meetings of the sub-teams. He or she may review the agendas and minutes of the meetings and make recommendations.

Spokesperson

The pandemic spokesperson during the remediation phase of pandemic or crisis management periodically conducts press conferences, bringing the media up to date on progress during this phase. He or she will present and ensure that messages are delivered. If specific interviews with people involved in the pandemic or crisis are requested by the media and granted, the spokesperson must be careful to ensure that any organization person being interviewed other than the spokesperson is appropriately trained in addressing the media.

The following are the roles and responsibilities of the sub-teams.

Functional sub-teams at remediation and recovery

Legal. The legal sub-team is responsible for protecting the organization from litigation associated with the pandemic and the spin-off crisis management team activities during and post-pandemic. They also work to eliminate legal delays in satisfying the pandemic and crisis management sub-teams, addressing noncompliance situations. The legal sub-team works with insurance groups and helps respond to financial risks because of the pandemic and the crises post-stability and reviews possible legal issues facing the organization because of the pandemic. The legal sub-team has the ear of the organization leader and addresses possible suits. When the work of the legal sub-team is complete, the sub-team is then sunset.

Human resources. The human resources sub-team focuses and acts on the needs of those impacted by the pandemic and spin-off crises. The sub-team may be involved in loss of shelter, jobs, and income. At the conclusion of their effort, the sub-team sunsets.

Communications. The communication team shares accurate information about the virus, its variants, treatments, vaccines, hospitalizations, deaths, and much more. The issue today is not about what is being shared and more about what is believed. There is a classic curve that explains this believability issue (figure 3). Once the communication of science crosses the disbelieve line, it is very difficult to change the thinking of people no longer believing in science. There are many reasons for this. One is misinformation that goes unchallenged. If it remains unchallenged, it becomes incorrect information, but it is believable to those who choose to believe it.

This communications group ensures that information is effectively shared among all those directly and indirectly involved in the pandemic and spin-off crises.

The role of the communications sub-team post-stability is to communicate to the media the next steps concerning remediation and recovery. Further, the communications sub-team provides only facts and not opinion. This will become critical as interviews take place with the media where the media would like opinions such as, why do you think it happened? Was the response time adequate? Could you have done something different to prevent deaths? The communications sub-team at remediation provides written and verbal messages and demonstrates a clear, positive effort to support the affected community. It is important that the team be open and honest. Tell the truth! Sincerity and believability are the key to all messages, and the messages must be perceived as real and meaningful to be effective. The role of the communications sub-team at recovery is to create messages, seek approval of the pandemic and crisis managers and, as appropriate, the pandemic or crisis management teams, and issue statements, written and verbal, to the media, announcing that recovery has been achieved. It identifies, with total sincerity, the sadness about those who have died because of the pandemic. Further, it commits the organization to work to ensure such a pandemic will never happen again. At the conclusion of their effort, the sub-team sunsets.

Information technology. This group ensures that all computer needs are addressed during the pandemic and through all activities of the remediation and recovery stages. At stability, the information technology sub-team checks into systems involved in response to the pandemic to determine if they are functioning properly. When some systems are not functioning, the team works to ensure backup systems are functioning. The role of the information technology sub-team at remediation is to ensure that what computer issues need to be fixed are fixed as soon as possible. The sub-team at recovery ensures all systems are functioning properly. Further, the team provides a report to the pandemic and crisis managers; after which the sub-team sunsets.

Possible additional sub-teams

Medical. The medical group gives guidance and medical support during the pandemic. The sub-team will also provide guidance to other health providers working with those impacted by the pandemic and spin-off crises. At the conclusion of their effort, the sub-team sunsets. Science and medical experts should avoid interviews with the media because the agenda of the media is not the same as the expert.

Recovery. Recovery, as it relates to pandemic and crisis management, strives to bring a return to normalcy, where normalcy is defined as the condition that existed just prior to the pandemic. It is critical that everything be brought back to a point as close as possible to where it was an instant before the pandemic was declared by the WHO. It must be remembered that people are working through their losses as best they can, and as difficult as it may be, there is a need to look to the future. While the crisis sub-teams work to bring about recovery, it is unlikely that the condition prior to the pandemic will be fully reached; however, the closer the sub-teams move to the condition similar to that just prior to the pandemic, the better.

The pandemic and crisis managers. The role of the pandemic and crisis manager at recovery is to ensure that the effort to bring about recovery, a returning to the condition an instant before the event, is as complete as possible. The pandemic and crisis managers are put in place when the pandemic or a spin-off crises from the pandemic reaches stability. With stability in place, the need to have the full pandemic management team or a full crisis management team in place is questionable. The pandemic and crisis managers communicate regularly with leadership of the pandemic and spin-off crisis management team leadership via written and verbal reports. The pandemic and crisis managers may call the pandemic or specific crisis management teams back into session if they believe it appropriate.

The pandemic or spin-off crisis spokesperson. The crisis spokesperson at recovery announces to the media that recovery has been completed, but they shall never reach total recovery because of those who were lost. The announcement should, if possible, ensure that measures have been put in place to hopefully prevent such a pandemic from ever occurring again.

Closure. As with any tragedy where lives are lost and those who remain are deeply affected, time is the healer, but closure is difficult. Those who survived the pandemic will never forget. The pandemic and all its learning must be documented not only for legal reasons but also that the learnings may be shared with others around the world. Every effort must be made to share the information from the pandemic in order to prevent such an occurrence from ever happening again. Closure is an attempt to move forward and try to put the pandemic and all its impacts behind as a learning and to put in place measures to prevent the event and subsequent impacting issues from ever occurring again.

CHAPTER 16

The Local Community, Country, and World Community and a Pandemic Situation

The community and a pandemic

As with any natural disaster or upset condition, the concerns must be with those that have been impacted by the pandemic. The first concern is the impact on the lives of the members of the community experiencing the pandemic.

Specific concerns must be addressed to those identified as critical receptors. Generally, these are the aged, the young, and the infirm. History tells us that these are the people who are most impacted during a pandemic or spin-off crisis situation: the aged because many are already suffering from illness or their health is compromised; the young because a pandemic illness could affect the rest of their lives; and the infirm because these people are ill, compromised and in need of care, and susceptible to further illness. If someone who is infirm, aged, or both is displaced during a pandemic, it may be critical that they have their medication or access to their medication should they be quarantined in their homes, or from wherever they reside.

Governments must be prepared to receive phone calls from local hospitals looking for guidance since they are having an influx of people experiencing respiratory issues from COVID-19. The caller from the hospital should be put in touch with experts for guidance. This could be industrial hygienists, physicians, and virologists to name only three. The industrial hygienist can provide guidance on protective exposure equipment, and the physician can speak directly with the hospital emergency room physician and provide guidance concerning treatment of affected persons, and the virologist can provide information about the virus involved.

Communities are usually prepared for events such as tornadoes and hurricanes, especially if there is a history of such occurrences; however, they are far less prepared for a pandemic. Community preparedness and first-responder training are critical to saving lives during a pandemic. Government operations must interact with the local community first responders where they exist. This can be in the form of training, vaccine creation and distribution. Government facilities must coordinate efforts with local community leaders. The community needs to see and understand the efforts underway to protect them during a pandemic. It is highly recommended that pandemic exercises or drills be conducted and involve community resources, police, fire, ambulances, and local hospitals. The drills will ensure the community that local services are prepared to address a pandemic. Management skill sets such as organizational, interpersonal, and leadership must be in place and working when a pandemic occurs. Leadership must be clear and apparent in the community. All levels of government must be seen as working together to bring the pandemic and spin-off crises to stability as quickly as possible with minimal impact. Leadership must be presented in a calm and deliberate way such that the community is reassured that every effort to minimize the impact of the pandemic is taking place and efforts to protect the health and welfare of the community are in place. The health and welfare of all involved must be foremost in the minds of the leadership and all levels of government and its support services.

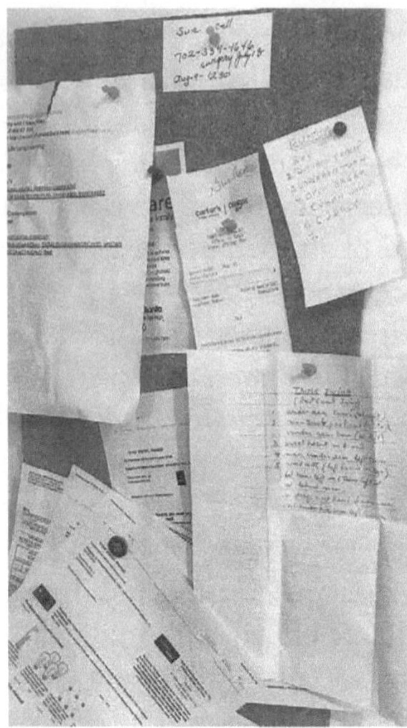

Messages

As with any pandemic or spin-off crisis situations, communication is critical. The people in the community must be kept informed as to what is occurring to minimize the impact and maintain control of the situation. The communications, which, if possible, should be joint, must come from community leadership (the mayor, police chief, among others) and the leadership of the federal government. Communications must be specific and done in a timely manner. Each time messages are delivered to the community, questions should be taken from the media. The responses must be clear and to the point. Messages that are important for the community to hear may not only be presented at the outset of a press conference but also should be restated, where possible, as part of responses to questions. The individual who responds to the media, the spokesperson, must be trained on how to speak to the media, must be empathetic and sympathetic to the families of those who have died because of the pandemic, and they must not stonewall a response in any way. If the spokesperson does not know the answer to a question asked by the media, he or she needs to explain to the person who asked the question that he or she will gather the information to provide an answer as soon as possible. The answer should be provided at the next press conference or earlier. People's fears must be addressed along with their concerns. Please see the chapter on media, which will provide additional information.

The establishment of a level of trust is critical to the success of satisfying the issues caused by the pandemic. The trust is created over time so that when a pandemic takes place, the communication of information to the community, the first responders, police, and fire departments can be trusted.

Community fire, police, and ambulance services are typically well trained and knowledgeable in addressing emergency and crisis situations. Oftentimes, if the upset, emergency, or crisis occurs because of a derailed train or an overturned eighteen-wheeler truck, community services will be the first on the scene to establish a command center to address the situation. A pandemic is different. Specific training of the first responders, police, fire departments, and the like will be necessary to successfully address a pandemic.

Remember, at a time of pandemic and spin-off crises that impact a community, there will be great fear and concern on the part of the community. They will fear for themselves, their loved ones, and especially their children. They understand that this sort of thing can and does happen, and they will depend on people with knowledge, skills, and abilities to tell them the truth and what to do and to remedy this situation as soon as possible.

They do not want to become ill because of the pandemic nor do they want an illness to come to any of their loved ones. You must work with all involved to communicate information, minimize the impact, bring about stability, and assist in the remediation and recovery efforts. If you are the spokesperson, you must deliver the messages and, if possible, provide assurances to the community that all levels of government are working together to minimize impact and bring this pandemic to stability as soon as possible.

While you will be making every effort to save lives, minimize impact, and bring the event to stability as quickly as possible, you must be careful to stay with the facts. Do not offer opinion even when asked. Who or what is responsible for the pandemic, epidemic, or crisis is not a subject for discussion during the time of minimizing the impact and bringing the pandemic, epidemic, or crisis to stability. If asked how this could happen, the recommended response is, "I do not know the answer to that question, and I am sure it will be addressed later. For now, our concentrated effort is on saving lives first, aiding families that have lost loved ones, and to address issues where people have lost homes and other life sustaining items." At no time place blame or take responsibility for a pandemic, epidemic, or crisis. While not always possible, where it is possible, rapid action will minimize the impact on the community. Everything possible must be done at the outset of a pandemic to try and minimize spin-off crises. As with any pandemic where there is a potential for people to become ill and die, there is fear on the part of the families that loved ones will become ill and die. While management needs to deliver messages as to progress being made to bring the pandemic to stability, messages to the community must reflect the loss of those who have died because of the pandemic in meaningful, sincere, sympathetic, and empathetic terms.

CHAPTER 17

The Media and the Crisis Situation Rooms

The news media working on a pandemic that may lead to many crises will be reporting on many levels. There is television, both network and cable; the Internet, which reports news and permits almost instant global communication of the pandemic; the print media in the form of newspapers and magazines; and radio, both local and satellite, that will report the pandemic and the activities surrounding the pandemic. The way people get their news has changed over the years. Today television and the Internet are the primary ways news is acquired. Newspapers, while they exist, are not typically the media of choice to receive news; indeed, some newspapers are either out of business or struggling and having difficulty surviving. One of the primary reasons is that the news of today becomes worldwide almost instantaneously. While in the past people waited for their daily newspaper to read what happened yesterday, today the Internet, radio, and television will report events within minutes of occurrence. Some media monitor local 911 calls to be on top of events when they occur. An organization having the potential to be in crisis, this includes the government, needs to understand the objectives of the news media at times of a pandemic or spin-off crisis and government's objectives during crisis and the difference between the two. It is also important that government leadership respond, not react, accordingly when faced with a pandemic that may lead to crises. In other words, the bottom-line message for the government is, "Be prepared."

The objectives of the media are essentially around gathering information, the drafting of the story, providing news and information to the public, and keeping the television watchers, radio listeners, and readers of newspapers and magazines interested in a story. At times the media may make a story controversial, which results in watchers, listeners, or readers wanting more information. There is a reporting of facts, opinion, and what might have gone wrong; and there is also an attempt to get the story behind the story. The objectives of government having experienced a pandemic that has led to crises is to have the crises brought to stability as soon as possible and for the news media to report the facts. There is a hope that the media will be fair and sensitive. Most, if not all, governments want to answer rational questions to the best of their ability, and they want to be honest and truthful. Anything else too often leads to a lack of trust and serious problems for the government experiencing the pandemic or spin-off crises.

Consider the media arriving at a government-operated nursing home experiencing a high level of COVID-19 infections and high death rates. What does the organization do if anything? There is a set of fundamentals that should be followed, understanding that the objectives of the media in coming to the scene are not necessarily the same objectives of the government. Understand that the media needs you more than you need them for them to provide the story to the public. Under no circumstances stonewall in providing information, or the media will write the story without your input. Remember, you do not debate the media in the media because they will have the last word. It is important to get your standby statement ready and provided to the media. An example of what not to do concerning the media during a pandemic or spin-off crises follows.

The media have arrived at the scene. They are in the parking lot of the facility with cameras rolling, and they are taking pictures. They arrive at the front door with their cameras rolling and request entry to interview residents. The media stops the first person they see before they arrive at the front door and asks them, with microphone in hand and cameras rolling, "Why do you believe this place has such a high infection rate? What do you think happened?" The startled employee is running to their car to go home and responds to the question with, "I just found out about the large number of COVID-19 cases, and I am frightened. I do not think we have enough masks, and they are not yet offering vaccinations." The reporter asks, "Who is responsible for this?" The employee responds, "I am not sure, but I do not believe I am coming back."

The reporter turns to the camera and says, "You heard it. Management from this facility hid the unhealthful conditions from residents and employees, possibly causing the death of many people." The reporter reaches a member of management and states, "One of your employees running for their life said that you do not offer vaccines and do not have enough masks. Why did you decide no masks and did not offer vaccines? How do you feel about killing people, and what else are you keeping from your employees, perhaps exposure to toxic chemicals and carcinogens? How do you feel about making people live and work at a cancer-causing facility and has a federal Occupational Health and Safety Administration (OSHA) audit, or perhaps the state or local government ordered your facility to correct a situation because you are not in compliance?"

In response, the management individual on camera says to the reporter, "I do not know what you are talking about, and who let you into this facility? You are in danger if you stay here. I am needed elsewhere. Please leave immediately." The response of the reporter on camera once the management personnel has left the scene is as follows: "Management refused to speak with us and asked us to leave the property. Management did not deny that they have been hiding the fact that this is an unsafe operation. Further, management did not deny that it has kept the fact that this is an unsafe operation from their employees, and the management representative also did not deny that there are chemicals on-site that will cause cancer."

In the example given above, the first person that was stopped by the media should not have answered any questions; he or she should have referred the media to management for answers to the questions. In addition, the individual who answered the questions should have said, "You do not belong here. Come with me. You need to leave immediately." Where possible, the employee should have sought out a management person to escort the media out of the building and told them to leave for their own health and safety. If they refuse to leave, police should be called to escort them out of the building. It is important to realize that a pandemic that results in crises not only puts employees, the residents of the nursing home, and the media at risk but the community as well.

Employees and management should be aware of the procedures concerning media being on-site, which must be implemented at the time of the pandemic. The media will only get in the way of the needed activities involving saving lives, which includes the lives of the media and others. The media needs to understand that their health and safety is at risk, and they should not enter a facility experiencing a pandemic or crisis. No one asked the reporter or the cameraman if they were vaccinated and where their masks were.

Not all countries are doing the same thing in managing COVID-19 and its variants. Many countries around the world have approached the management of the pandemic differently. Some in the United States do not believe the government has the right to tell them to do anything. Therefore, mandates are difficult. However, companies can mandate employees to be vaccinated, and many have done so. This will move the effort closer to herd immunity. Many states have passed laws that prevent local government mandating vaccinations and mask-wearing. While in Europe some countries are requiring proof of vaccination before entering a restaurant. It appears that while actions on the part of Europe meet the guidelines to manage a pandemic, the United States action may prolong the pandemic.

There is an appropriate way to address the media during a pandemic. The media must be managed, and they are not in charge! For example: at the location where the pandemic is taking place such as a nursing home, block the parking lot to all vehicles except emergency vehicles, and there should be no pictures permitted closer than one thousand yards from the nursing home. This is both for the health and safety of the residents, families of the residence, and the media. Zoom meetings may be arranged for families. In addition, picture-taking is forbidden within the facility again for proprietary reasons. Cameras must be left in vehicles. The media may be permitted inside the facility post the pandemic; however, access to the facility may be granted if and only if a COVID-19 test is taken and passed by those who choose to enter. Vaccinated documentation will also permit entrance. The exception is for police, fire, and ambulances and other emergency vehicles engaged in activities of saving lives and putting out fires.

During the media briefing, a standby statement will be provided. The media will be told that there will be a press conference, and the media should be told when the press conference will take place. They should also be told that questions will be permitted at the press conference. Regarding any quotes in reporting comments from a spokesperson or anyone else in the organization, they are to be shared with the spokesperson for approval prior to publication. Comments on any report should not be made to the media unless and until the media provides the report forty-eight hours prior to questioning. Remember the media needs you more than you need the media. Also remember that nothing is off the record. If in discussing an event with the media prior to an interview, the reporter says, "Off the record, why did you wait before responding to the pandemic?" Remember there are no "off-the-record questions." Do not let your guard down. Answer all questions on the record. It is also important to remember that you are in charge when asked to provide information to the media regarding your response to the pandemic and the spin-off crises that occur. An interview needs to be requested by the media, and parameters for the interview should be established. If the parameters for the interview are not met by the media, feel free to cancel or postpone the interview. You will have an opportunity to explain that the media was not prepared for the interview and you postponed meeting with them. Explain that you hope the media is prepared soon since you are ready to discuss the pandemic with them.

You should identify where you want the interview to take place and ask for the names and number of media who will be asking questions, preferably one. It is appropriate to ask for questions in advance. If it is not a live television or radio show, request a copy of all video tapes of the interview before they are aired. It is also important to limit the length of the interview. If the interview will be a video to be shown on the news, it is important that you simultaneously make a video of the interview. This prevents selective cuts of the videotape from being shown; it should also prevent text from being taken out of context and prevents unnecessary embarrassment. It is also appropriate to have a member of your communications staff present off camera during the interview. This is especially important if the interview is being taped. The communications person can stop the taped interview if he or she judges that the questions are not in line with discussion of the pandemic that led to a spin-off crisis. For example: if the interviewer begins to speak about a topic other than the pandemic, the interview should be stopped by the communications person or the person being interviewed. If the interview is live, it can be explained ahead of time that questions concerning topics other than the pandemic and spin-off crises are not a subject for this interview. If this is disregarded and a question on the economy is raised during a live interview, it is appropriate to say, "This is a subject for a different interview. Let us stay with the issue of today, pandemic and spin-off crisis we are faced with, and how they are being addressed. I am sure those watching this program are more interested in the pandemic and spin-off crisis than they these other topics." It should be a given that the media realizes that satisfying this pandemic or spin-off crisis is far more important than another topic.

It is critical during the interview that the spokesperson being interviewed is prepared with messages. The messages should be limited to three to five as a maximum. When the spokesperson receives questions from the reporters at the press conference, keep the messages you wish to deliver in mind. Answer the question from the reporter as correctly and accurately as possible, and without opinion. Immediately after answering the question, transition and deliver the desired message or messages.

For example: let us say the reporter asks, "Were you aware of the need for a new pandemic response plan?" The response to the reporter should be as follows: "We do not believe that our pandemic response plan is in need of renewal. It did need to be updated from our last pandemic exercise, and that was done. We recently learned that our efforts involving COVID-19 testing, mask-wearing, and vaccinations are the best in the world."

Taking a closer look at what was stated above, the response to the question from the reporter was, "We do not believe we needed to renew the pandemic plan, only to include the learnings from the last exercise and this was done." The message that was delivered without taking a breath, "We recently learned that our efforts involving COVID-19 testing, mask-wearing, and vaccinations are the best in the world."

In responding to questions from the media, be sure and deliver your answer in a deliberate manner no matter the question. It is critical not to react to the question. The next step downward from being reactionary is defensiveness. When you become defensive, the party is over. Whatever is said thereafter will be ignored. The focus will be on the defensive posture of the person being interviewed. The following are examples of some messages for the media, assuming again these are correct and therefore can be said:

"We are working with nursing homes to ensure COVID-19 testing."

"We are acquiring an ample quantity of vaccines."

"We are expanding hospital capacity."

"We are using every available resource to satisfy this pandemic and spin-off crises."

It is highly recommended that individuals who will speak to the media be trained in how to respond to questions from the media. Many companies provide such training. The trainers are usually retired media people who give insight as to what should be expected from a media interviewer during a pandemic or crisis. Do not let your guard down with media comments such as, "We are having technical problems with the video recording equipment," and while you are waiting for the equipment to be repaired, which is not broken, a question may be asked where you provide an unguarded response that appears on the 6:00 p.m. news. Do not get flustered. Many times, reporters know about the person they are interviewing and will not hesitate to bring up an embarrassing experience from the past of the person being interviewed. In addition, do not stonewall. Do not delay your response. If you do not have the answers, say, "I do not know the answer to that question," but quickly add that you will get the information concerning the question to person asking the question as soon as possible. To establish and maintain credibility, make sure you get back to the person with the answer. Stay factual and do not give opinion.

The issuing of the standby statement with the date and time for a press conference is not stonewalling and provides minimal information regarding the pandemic and its spin-off crises. If you stonewall the media, they will write and report something, and it will probably not be to your liking. Provide information outside the standby statement that has been reviewed by the sub-team and, hopefully, time permitting, approved by the pandemic and crisis management teams. A classic example of media not being managed and creating news include magazine-format television programs. One of these programs in reporting on a construction of a new nuclear power plant included an interview with a nuclear engineer who was fired by the plant manager, and it also included a review of a line drawing of the plant piping. The saving grace was that the company building the nuclear plant filmed everything that was filmed by the magazine-format television program. This simultaneous filming by the company and the television magazine program clearly demonstrated that important information was left out of the television program that was aired, and the magazine-format television program did not check sources before they aired their report. It was learned later that the fired employee was not an engineer and did not have a degree. The on-camera question involving safety was cut when the company person responded to a question with, "No, we do not include safety measures at this point in the process." The part of the interview that was not included was the next sentence, where the person answering the question said, "It would be wrong to include safety measures at that point in the process. We employ them here"—as he pointed to the line drawing—"where it provides a much-safer environment."

When considering being interviewed by the media regarding a pandemic or spin-off crises, you must reflect on the media's objectives and know they are different from your objectives. There is need for training before addressing the media and a need to have the media meet your requirements before granting an interview. All this comes under the label of managing the media. While some may say that it is impossible to manage the media, that is not correct. Managing the media is not an attempt to alter facts or lighten the impact of a pandemic or spin-off crises. It is an attempt to achieve balanced and fair reporting.

Equally important is the situation room where the crisis is managed. The room should be designed such that all individuals working to satisfy the pandemic or spin-off crises can work together, and it must also permit individuals to work in small groups and, as necessary, to work alone. In the latter case, this would include the making of important phone calls to other government officials. The room should be designed in a manner that fosters communication among all in the room and with small groups working on specific issues such that minimal disruption occurs among those working together. The room must enable a government to walk the talk, enable a strong working relationship with stakeholders, and maintain or enhance the acceptance of government experts working on the pandemic. This is no small order.

The room should foster the commitment to save lives first. It should permit the accurate monitoring of the pandemic and spin-off crises, response activities, and the assigning and acquiring of needed resources. The room configuration should help in the building of confidence as people share in the activities to satisfy the pandemic and crises and thereby ease the minds of all involved. The room in its configuration should also ensure the accuracy of information exchange, and beyond saving lives first and caring for those with COVID-19, it should foster the making of informed decisions.

Situation room

- Design
    o   Lecture hall design
    o   Mission control
    o   Boardroom

o   Multiple tables in same room addressing different spin-off crises from the pandemic and the pandemic management team

There are many choices of design for the room. It could be set up as a lecture hall, a marketplace, mission control, a boardroom, and many other configurations. If a lecture hall is chosen, the room is set up as a formal room for a lecture. Chairs may be in a semicircle, and the leader in front of the room does most of the speaking. This is more of a command-and-control effort with limited interaction. In the marketplace model, each of the main functioning pandemic and crises areas are in the room at separate working tables. This would include perhaps human resources, health, and safety, information technology, and legal, among others. In this model, there is good interaction within functioning groups; however, there is limited interaction among the groups present in the room.

Mission control is a setup much like that of the National Aeronautics and Space Administration (NASA) configuration set up for launching of satellites and putting people into space. Those present are individuals at computers doing their job. The leader is in the room but has limited interaction with those at the computers. Information flow among individuals is limited. In the boardroom configuration, key players interact with one another. Actions such as monitoring the media takes place outside the boardroom. The leader may ask questions, and everyone in the room can clearly hear him or her. Feedback from involved parties is immediate. However, one must go outside the room or invite people into the boardroom for periodic updates.

It is obvious that one design does not fit all. As part of the pandemic and crises management plans, use one of the many decision analyses techniques. For example, Kepner-Tregoe decision analysis or some other decision analysis may be used to select the appropriate situation-room design for your organization to best manage a pandemic or spin-off crisis. Likely, the room design will be a combination of those identified above. If a pros-and-cons approach is taken, there will be a need to make a list of what works and what does not work. Other analyses result in a listing of what is absolutely needed and what is desired for a fully functioning pandemic and crises management situation room.

Aside from the room where the pandemic and crisis are being managed, there should be other rooms or locations that need to be considered. These other rooms, just off the situation room, are for contacts with people working in the spin-off crisis areas, a room to monitor the media, a room for backup equipment such as computers and telephones, and a refreshment room or food room. Depending on the length of time that the pandemic or spin-off crisis team will need to meet, there should be an area for, if needed, the setting up of cots and for sleeping. Considering the nature of the pandemic and crises and the discussions that will take place, there should be limited access to the situation room. A magnetic-card entry system should be considered. In addition, a room for discussions with stakeholders should be considered. There should be a room for interaction with the media or for press conferences. There should also be restrooms and areas with showers for extended stays. There should be small conference rooms, if needed, for sub-teams to meet. However, the creation of a building to address all the needs of the situation room is not necessary. There should be a review of conference rooms that are currently used for other reasons and surround a large centrally located board room. This arrangement should be considered for use during a pandemic or spin-off crises.

Aside from identifying a room where the pandemic or spin-off crises are to be managed, there are other needs. These include supplies such as pens, paper, markers, and the like. A cabinet or computer programs with state and local emergency response plans should be available. These will be used should information be needed concerning the location undergoing the spin-off crisis. In addition, three-by-five cards with the role of each of the crisis management core and ad hoc team members should also be in this cabinet so the team member roles may be satisfied in the absence of a member and their alternate at the time of a crisis. The crisis room should be equipped with Wi-Fi, a VCR or DVD to record media broadcasts, and must be hardwired for computer hookups. It should have secure

Internet access, televisions, telephones, the ability to scan documents into a computer, a copy machine, a fax machine, and printers.

In the area where the press will be meeting, there should be folding chairs for press conferences and a podium. In the situation room, there should be a whiteboard, a computer software system for the posting of activities underway so all may see progress, and an intercom system for general announcements. Where possible, communications via telephone, Internet, and the like should be established between the situation room and pandemic or crisis hot spot areas. If videoconferencing is available, it should be used so the pandemic or spinoff crisis management teams can see what is happening around the country and the world so they can respond accordingly. The pandemic and crises management teams need to access to internal and external databases, and this should be established by the internet technology group. For example: the pandemic and crises management team will need access to government statistics and reports for each state.

All these will help in the development of messages. Other databases that should be available to the pandemic and crises management teams include government health statistics, safety agency contacts with phone numbers, contacts, and phone numbers of local television stations, the national emergency response center, health and safety statistics. From these organizations, positions and public statements during the pandemic and crises will be available. For example (assuming the following is an accurate statement), "Our effort responding to the aids pandemic was highly effective."

Since spin-off crisis management rooms will hopefully have limited use, these rooms should be primarily used for other functions. This includes conference rooms dedicated for use by crisis sub-teams, and rooms that will house cots, food, and the like need to be identified in the pandemic and spin-off crisis management plans but are used primarily for other activities.

What has been presented here as a pandemic or spin-off crisis situation room is obviously an ideal situation. It is unlikely that many organizations will be able to dedicate rooms and commit resources without a pandemic or crises taking place. What has been described can be addressed in the pandemic and crisis management plans where many of the rooms, such as the pressroom, may be in another building from the situation room. Important here is to understand that all the identified rooms may be needed in total or in part at the time of a pandemic or spin-off crises.

What happens should, for any reason, the situation room not be available? An alternate location should be known by all pandemic and crisis management team members. If the situation room is not available at the time of a pandemic or spin-off crises, alternate locations should be selected such that the criteria of the situation room layout are met as much as possible.

In summary

The pandemic facilitator is leading the pandemic and crisis management teams throughout the plan. Decisions are made by the team to send internal or outside resources to assist in managing the pandemic or spin-off crises. The media are being monitored in another room by the communication sub-team to see what is being said about the pandemic and spin-off crises. The spin-off crises management teams are sub-teams as needed and provide updates to the pandemic management team. In time, the pandemic or crises management teams will call for the activation of the room where the media will gather for a press conference. At no time will the media be permitted near the situation room. As needed, the pandemic or crises management teams will call for the setting up of vending machines, food to be brought in from a caterer, the establishment of a room for eating (in some cases, this will be the organization's cafeteria), and the setting up of cots for sleeping. All the above will be initiated on an as needed basis. This should be explained in the pandemic or crisis management plans.

Consider auditing the ability to respond to a pandemic. Perhaps the best way to do this is to conduct a pandemic response exercise. This will help identify the shortcomings in the pandemic plan and the plan to address spin-off crises from the pandemic.

CHAPTER 18

The Pandemic, Crisis Scenario Creation, the Exercise, and the Critique

A pandemic scenario should be created and an exercise carried out to test your pandemic and spin-off crises management plans. It should be scripted so that people play roles, and the pandemic and crises management teams are interrupted with issues and situations while trying to manage the pandemic and spin-off crises. The exercise is made as real as possible by bringing in television cameras to the leadership and the conducting of interviews with team members. For the exercise, the television cameras should not come from local television stations; they should be brought in from local media training programs. At the conclusion of the exercise, have a critique concerning how the exercise was handled and to see if the plan was followed and worked as designed in total or in part. The follow-up from the critique usually manifests itself in improvements to both the pandemic and crises management plans. It is for this reason that the critique is very important. The team should also undergo training in addressing misinformation and conspiracy theorists.

Many questions surround the issue of scenario creation; for example, why is a pandemic scenario needed? What exactly is a pandemic scenario? Who should create it? Where and when should it be created, and who will participate? How do you create a mock scenario? Taking the questions one at a time, there is a need to develop a pandemic scenario so that the pandemic and spin-off crises team members may participate in an exercise so that when a pandemic occurs, members of the pandemic and crises management teams will have had the experience of managing a pandemic and crisis situations. It is for this reason that the crisis exercise experience must be made as real as possible. What is a crisis scenario? It is a document that is prepared as a scripted exercise, and it should be as intense and as real as an actual pandemic and crisis experience. Where should it be conducted? It should be conducted at the government location housing the largest number of people who would be involved should there be a pandemic and spin-off crises. When should the exercise take place? It should take place at least once per year to test the plans and to give the pandemic and crises management team members the experience without a pandemic or spin-off crises taking place.

Who should be involved? All the members of the pandemic and spin-off crises management teams and their alternates should be involved and should take on the scripted roles. Since a scenario is scripted, those who take on the roles will be prompted when to speak according to a scripted timetable. For example, eight minutes into the exercise, the pandemic management team learns that the local radio station is blaming the government for not preparing its citizens for a pandemic.

How is the exercise to be carried out?

The pandemic management team facilitator announces that there will be an exercise sometime over the next couple of months. In this way, the pandemic management team is not prepared for the exercise as they would not be prepared if a real pandemic occurred. The date and time of the exercise, not the content, should be shared only with the leader to assure that the exercise will not interfere with any important scheduled event. While the exercises are conducted during normal working hours, it is of value to hold at least one exercise after normal working hours. This will test the ability to reach the pandemic management team members and to gather them in the situation room during nonworking hours. Before the exercise begins, it is important to tell the pandemic management team that if a true pandemic were to be announced by the WHO while the exercise is underway, the exercise will be stopped, and the real pandemic will be addressed.

Once gathered, set the scene for the pandemic management team by describing what is happening in as much detail as possible even though the pandemic that is being described is not taking place since this is an exercise. It is important to engage all members of the pandemic management team by having them ask questions and answer as many questions as possible within the context of the script. The person taking the role of the pandemic facilitator reads the text received from the WHO that contains possible actions the team may choose to take. These include calling for quarantine if exposed to the virus, the social distancing of six feet, the wearing of masks and vaccination, if available, possible treatments for the virus and other possible measures. Once the exercise is complete, take a ten-minute break. After which undertake a critique of the exercise—what went right, what went wrong, and what was learned? The learnings from the critique and the overall exercise will be used to update the pandemic and spin-off crisis management plans.

Sample scripted exercise

This exercise does not reflect the actions taken by the United States Center for Disease Control (CDC) in responding to the WHO announcement of a global pandemic for COVID-19. It is simply a scrip denoting possible actions of a pandemic management team and the complexities of addressing a pandemic.

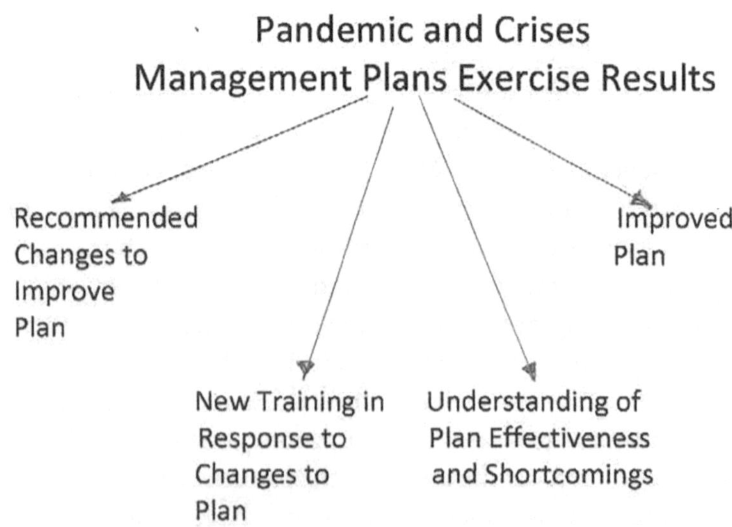

Scripted exercise

It is fall, peak season for viruses to exist and spread. There is general knowledge that a virus has appeared in Indonesia and appears to be making its way in hot spots in different parts of the world. However, the World Health Organization has not yet declared a global pandemic. You wake up to learn that late last night the spread of the virus in the world met the WHO criteria and the WHO has declared a global pandemic.

The following is an example of a scribe's notes from the exercise to test the pandemic management plan.

7:00 a.m. With a global pandemic declared, the facilitator arrives at the office and pulls out the pandemic management plan.

7:10 a.m. Other members of the pandemic management team begin calling to make sure the facilitator is aware of the WHO announcement.

7:15 a.m. The facilitator calls the individual in charge of communication and asks that they contact pandemic team members and ask that the be present in the situation room at 9:00 a.m.

7:45 a.m. The facilitator learns the names of the team members that will be coming to the situation room and those who will be participating virtually. Two team members out of the country, one in England and one in China, both will be via zoom.

8:00 a.m. The facilitator begins looking at the resource list of people available to help and to add to the team as ad hoc members.

8:15 a.m. The facilitator's administrative assistant enters with a list of phone calls from various media, TV, radio, cable news, and the print media, asking for an interview.

8:45 a.m. The facilitator picks up the copy of the plan and heads to the situation room.

8:55 a.m. Team members begin to arrive in the situation room.

9:03 a.m. All team members expected to be present have arrived, and those on virtual are present and accounted for.

9:04 a.m. Each team member introduces themselves, their position in the organization, and what they bring to the table to help satisfy the pandemic.

9:10 a.m. The facilitator with the approval of the leader and the pandemic management team assigns a scribe to take notes and keep a running list of team activities and the time they occurred.

9:15 a.m. The facilitator with approval of the leader and the other members of the pandemic management team assigns the pandemic management plan spokesperson and informs them of the requests for interviews. The facilitator's administrative assistant will provide the list of requests for statements and interviews to the spokesperson. The facilitator asks the spokesperson to contact those who have called and say that the US pandemic management team will be meeting later this morning and a statement will be available later this afternoon.

9:30 a.m. Each team member receives a copy of the WHO pandemic declaration, and the team spokesperson reads the declaration aloud so that everyone hears it together.

9:35 a.m. Discussion begins to determine if the WHO has identified the United States as a country is included in the declaration of a global pandemic. If not, discussion should begin to determine if the United States should use the plan criteria to determine if the team should declare a US pandemic at this time or wait for further information.

10:00 a.m. If it is decided to wait to declare a US pandemic, a new meeting date and time should be established. This could be only hours, or the team can begin going through the criteria or establish a group to review the criteria and come back within hours with a recommendation.

Assuming the team declares a US pandemic immediately, the following would take place. If it is hours later, the time below should be adjusted.

11:00 a.m. The facilitator introduces the need for a standby statement to be provided to the media later today and asks for comments as to what should be the key messages we want to present in the standby statement. After which, the facilitator asks the spokesperson to have two to three members of his or her department draft a standby statement and have it available for the team for review and approval by early afternoon.

11:15 a.m. The spokesperson states that he has contacted those who have requested an interview, and all are looking forward to comments from the pandemic management team.

12:00 noon. The facilitator opens a discussion of the WHO recommendations, such as social distancing, wearing of masks, quarantining if you were in the presence of someone who has the virus, not gathering in large groups, business closures, school closures, washing of hands, use of disinfectants and the like.

1:00 p.m. A team member asks for information on what was adopted and used in fighting the 1918 virus. This was judged to be a good idea but was tabled to continue discussion of which of or perhaps all the WHO recommendations should be adopted.

1:30 p.m. The facilitator calls for a decision on what WHO measures will be recommended for adoption by the CDC and implemented.

2:00 p.m. All the WHO recommendations are adopted by the committee and will be passed along to the CDC for adoption and implementation. It was further decided that there will be a need to educate the public on why these measures were adopted and why they should comply.

3:00 p.m. It was decided to work through lunch that was brought into the situation room.

3:30 p.m. A team member points out that with business shutdowns, many jobs would be lost, and the absence of income will have a great impact on our economy. The members of the pandemic management team, not wanting to move to this issue because it would result in a loss of focus from the pandemic, called for the creation of a pandemic spin-off crisis management team that would focus on the economic impact of the pandemic and periodically report back to the pandemic management team.

The pandemic management team also discussed who should be on this economic crisis management team, leadership, and the like. Other areas that could result the creation of spin-off crisis management teams include but are not limited to school closures and all the issues associated with loss of education and social impacts, hospitals operating at total capacity and with a lack of medical staffs, the need for a vaccine, the antivaccine groups delaying progress to overcome the virus, and many more. The formation of spin-off crisis teams will permit the pandemic management team to remain focused on the pandemic.

4:00 p.m. The communications group presents a draft standby statement for the pandemic management team to review.
5:00 p.m. The pandemic management team completes and approves the work on the standby statement. The spokesperson prepares to present it to the media at the briefing set for 7:00 p.m. today.

I shall stop here, but the pandemic management team is not finished their work. They will continue their work at 8:00 a.m. tomorrow, or if appropriate, they will work through the night. This provides a framework for the exercise.

Unfortunately, the politics of the day do not always match the science. This can and has been a problem for both the pandemic and spinoff crisis management teams.

The scenario selected

The following sample template format may be used to manage a pandemic and spin-off crises.

Sample Pandemic Exercise

Day	Time in Minutes From Start of Exercise	Pamdemic Team	Media	WHO	Politicians	CDC	Antivax	Crisis Management Team
1	0			Declares a Pandemic				
	60		Requests Interviews					
	120	Team Meets						
	180					Identifies Possible Vaccines		
	240						Announces will not take Vaccines	
	300		TV Program Blames CDC For Virus					
	360							Economic crisis team formed
	420				Politicans Challanging CDC Guidelines			

The scripted exercise continues with messages being brought into the situation room, announcing some of the items listed above at the time in minutes indicated from time 0, the start of the exercise. The numbers reflect the time since the pandemic was learned. The pandemic management team takes up each of the items above and determines action to be taken, if any. The leader of the pandemic management team is in charge. The facilitator's job is to make sure the plan is followed, e.g., presenting the resource list, making sure all members are represented or their alternates, and all items in the approved pandemic management plan are followed.

An example of a written message follows. These messages are handwritten and handed to the facilitator, who reads it to the pandemic management team. If a team member needs to leave the exercise (as part of the script) to make a phone call, they do not make the phone call since it is an exercise; however, they stay out of the situation room for a time period equal to the estimated time the phone call would take, and they return to the situation room. If the media asks for a meeting with the leader, this should not be granted! The spokesperson can speak to the media. The reason is that the leader is expected to know all the answers and may be embarrassed if he or she says they do not know the answer. The spokesperson can get away with saying, "I do not know that answer but will get it for you and be back to you with an answer ASAP."

The critique is not done to find fault with any member of the pandemic or crisis management teams. It is done so that the lessons learned may be captured and used to improve the pandemic and crises management plans.

Questions the critique needs to answer:

- Who are the audiences for the messages created?
- Were they the correct audiences?
- What are the objectives for the messages?
- Were they the right messages?
- What is the strategy for delivering the messages?
- Was it the best strategy?
- What other strategies should have been used?

- Should more than three to five sound bites or messages have been created?
- Who else should have delivered the messages other than the spokesperson?
- Should the leader have presented the messages?
- Was there enough information to recommend that an economic crisis be declared?
- Should the messages portray confidence in the decisions and the direction of the pandemic management team?
- Should protecting the image of the CDC be a focus of the pandemic management team?
- In addition to the spokesperson, who should receive media training to speak to the media, and why?
- What was your overall impression of the exercise?
- What went right?
- What went wrong?
- What did you learn?
- How did the team do in responding to the pandemic and spin off crises?
- Was enough information provided to make sound judgments and informed decisions?
- Were you, as a member of the pandemic or crises management teams, satisfied with the identification of audiences, objectives, strategy, and messages that were drafted, and if not, how could have they been improved?
- Could the spin-off crisis management team have provided additional support during the pandemic, and if yes, what would it have been?
- How well did the team respond to information coming into the situation room?
- How might the plans be improved?

Key Questions to Ask of the Pandemic/Crises Management Teams Post Exercise
What Went Right?
What Went Wrong
What Did You Learn?

Will the COVID-19 pandemic ever end?

I have been asked the above question many times, along with, "Will we ever return to normal?" I shall take these two questions separately. The answer to the question, "Will this pandemic ever end?" is yes and no. Did the pandemic of 1918 ever end? Yes and no. As with the 1918 pandemic, the influenza cases never ended, but the number of cases dropped significantly. As for the COVID-19 virus, once the World Health Organization's definition of a pandemic is no longer met, the word "pandemic" will no longer be used to describe the situation. However, the following very important information will shed some light on the future of COVID-19 in the United States.

From October 1, 2019, to April 4, 2020, there were between 410,000 and 740,000 influenza hospitalizations, there were 39,000,000 to 56,000,000 global flu cases, and between 24,000 to 62,000 deaths. These numbers are primarily based on lab tests and modeling. Nevertheless, the important point is that these numbers reported over the six-month flu season did not make headlines like COVID-19 is and will continue making headlines. So, if you step back and consider the present number of COVID-19 hospitalizations, cases, and deaths, if they reduce to numbers similar to influenza in 2019 and 2020, reporting of COVID-19 should reduce significantly.

How do we get there? Each day that passes the number of people who get vaccinated rises, not by numbers they once were, but they rise. In addition, children between five and eleven are now receiving vaccinations. This could amount to as many as 28,000,000 more vaccinated. Once these children are vaccinated, and if those who refused to get vaccinated, regardless of the reason, get vaccinated, and if we include those who have survived COVID-19 and have developed antibodies, we should be very close to reaching herd immunity. If this occurs, we should be well on our way to returning to a lifestyle we enjoyed pre-COVID-19. Without this happening, progress in the United States will be much slower, perhaps taking through 2022 and well into 2023 and possibly 2024 before we reach herd immunity.

However, we cannot forget the rest of the world and especially underdeveloped countries. We may be looking at 2024, 2025, or 2026 for a global vaccination similar to smallpox or polio. This will also continue to impact global supply chains since we are very much a global economy.

Further, the United States deaths from COVID-19 recently surpassed the number of United States deaths from influenza in 1918–1919. It must be remembered that the population of the United States in 1918 was about 103,000,000, or less than a third of the population of the United States today. Therefore, the influenza pandemic was far worse than what we have and continue to experience in the United States from COVID-19. While we have a vaccine for COVID-19, we now also have a vaccine for influenza. Both are effective and will improve the health and well-being of the United States and world population. Recently COVID cases in Europe have surged. They are facing a new wave. There is a high probability that what is now going on in Europe will arrive in the United States in a matter of weeks. We are not yet where we all want to be, but we shall get there. We must be patient and above all be vigilant in fighting this virus.

Questions at Time of a Pandemic

*List of questions to be asked of locations
experiencing a pandemic or spin-off crises*

The following is an example of a wallet-sized card containing a list of questions that should be asked of the location calling in announcing a cluster of COVID-19 cases that may lead to an epidemic or a pandemic. These questions are important since the person calling in with the information is more than likely upset; therefore, it is best to ask specific questions to help gather information. The following are examples of some questions:

- What is the date and time the cluster was reported?
- What time was the call received?
- What is the name of the person taking the call?
- What is the name and phone number of the person calling?
- How many people have been infected with COVID-19?
- What is the name of the hospital or hospitals where those who are ill are located?
- Does the area reporting the clusters need medical help?
- Are media covering the situation? If yes, are they from radio, television, magazines, the press?
- Is the person reporting the cluster in need of a standby statement to satisfy the media?

- Please provide the names and phone numbers of all government agencies contacted regarding the illness clusters.
- Other questions as appropriate.

Note: This form, in part and with some modification, may be used during conversations with the World Health Organization.

APPENDIX 2

List of Resources

List of internal and external resources to address a pandemic and its spin-off crises

The following is an example of a wallet-size card for making contact with organization internal and external resources at time of pandemic and spin-off crisis. What follows does not include names and phone numbers. Your wallet-size card should include that information.

Note: Int—internal
Ext—external
X—names, locations, and numbers to be added
More may be added than listed here

| Area of | | Location | | | | Phone Numbers | | |
Position	Expertise	Int	Ext	Name	Office	Home	Cell
XXXXX	XXXXX	XX		XXXX	XXXX	XXXX	XXXX
XXXXX	XXXXX	XX		XXXX	XXXX	XXXX	XXXX
XXXXX	XXXXX	XX		XXXX	XXXX	XXXX	XXXX
XXXXX	XXXXX	XX		XXXX	XXXX	XXXX	XXXX
XXXXX	XXXXX	XX		XXXX	XXXX	XXXX	XXXX
XXXXX	XXXXX	XX		XXXX	XXXX	XXXX	XXXX
XXXXX	XXXXX	XX		XXXX	XXXX	XXXX	XXXX
XXXXX	XXXXX	XX		XXXX	XXXX	XXXX	XXXX

XXXXX	XXXXX		XX	XXXX	XXXX	XXXX	XXXX
XXXXX	XXXXX		XX	XXXX	XXXX	XXXX	XXXX
XXXXX	XXXXX		XX	XXXX	XXXX	XXXX	XXXX
XXXXX	XXXXX		XX	XXXX	XXXX	XXXX	XXXX
XXXXX	XXXXX		XX	XXXX	XXXX	XXXX	XXXX
XXXXX	XXXXX		XX	XXXX	XXXX	XXXX	XXXX
XXXXX	XXXXX		XX	XXXX	XXXX	XXXX	XXXX

APPENDIX 3

———

Sample Standby Statement

To appear on organization letterhead: this standby statement is read to the media by the organization's spokesperson as part of a briefing. After which, the statement is handed to the media.

Date

Good morning, afternoon, or evening (depending on the time that the statement is released).

______________________________ (the organization name) is deeply saddened (may use a different term if appropriate) that clusters of COVID-19 illnesses have been identified in numerous parts of our country (provide names of locations reporting large clusters of COVID-19 cases). At this point, we are aware of ____________ hospitalizations and ______________ deaths (provide accurate documentable numbers). Our hearts go out to the families of those who have died and those who are hospitalized. We are making every effort to bring all necessary resources to those in need in response to learning of these clusters. On the scene of the clusters are local community first responders, local fire and police, clergy, and counselors. We are sending internal and external experts to aid the communities that have been impacted. We have been and remain in touch with the World Health Organization who are providing guidance for managing a pandemic.

I shall return with an update of the situation in approximately four hours (if four hours is too long, identify a shorter time period) when we shall conduct a press conference where you may ask questions. I know you have questions now, so I shall try to answer only three of them, since I must return to the situation room, where I and other members of the pandemic management team are addressing the situation. I and members of the pandemic management team appreciate your understanding and patience.

Thank you.

APPENDIX 4

Pandemic and Spin-off Crises Management Plans

Pandemic and spin-off sub-crises management plans template

The following maybe used as a template for both pandemic and spin-off crises management.

What follows is a pandemic and crisis management plan template that may be used to prepare a pandemic and/or a crisis management plan or to modify existing plans. Please fill in the blanks below so it best fits your organization.

The template

Cover page: The pandemic or crisis management plan for (your organization's name here).

If your organization has a logo, it should be placed here on the first page.

* * *

Page 2 contains the following:
Date of last update ______

Plan identification, e.g., pandemic management plan or a specific crisis management plan. The specific crisis management plan needs to be identified by name. Example: The Supply Chain Crisis Management Plan.

Originally written by _______________________________________

Date: ____

Date when the original plan was approved: ___________________

* * *

Page 3 contains the following:

Revisions:

Revision number _______ written by _______ date approved _______

Revision number _______ written by _______ date approved _______

Revision number _______ written by _______ date approved _______

Revision number _______ written by _______ date approved _______

Organization's document control number: ____________________

Distribution list and approval:

Title   approved by   date approved   (signature)

____

____

____

____

____

____

____

* * *

Page 5 contains the following:

Table of contents

Item   Page number

Purpose:

Scope:

References:

Definitions:
Titles and responsibilities of core and ad hoc members:
Procedure:
Records:
Training:
Appendices:
Document management:
Pages continue
Purpose:

(The suggested wording from a previous chapter may be adopted, or please use your own wording.)

____

____

____

____

____

____

____

Scope:

(The suggested wording from a previous chapter may be adopted, or please use your own wording.)

____

____

____

____

____

____

References:

(The suggested wording from a previous chapter may be adopted, or please use your own wording.)

____

____

____

____
____
____
____

Definitions:

(You are encouraged to write in your organization's definitions of the following terms. Fill in below the wording you have selected.)

Crisis:

____
____
____
____
____
____
____

Event:

____
____
____
____
____
____
____

Upset:

____
____
____
____
____
____
____

Emergency:

Records:

Training:

Pandemic or crisis management flow chart:

Emergency:

Communications center:

____

____

____

____

____

____

Pandemic management and crisis management resource (list this is an appendix to the plan):

____

____

____

____

____

____

____

Management system—plan, do, check, act:

____

____

____

____

____

____

Situation or pandemic or crisis management room:

____

____

____

____

____

____

Assessment of impact:

____

____

____

____

____

____

____

Remediation and cleanup:

____

____

____

____

____

____

____

Recovery:

____

____

____

____

____

____

Closure:

____

____

____

____

____

____

Additional terms may be added.

Responsibilities of members of the pandemic and/or crisis management teams. This may include ad hoc members.

Please fill in the blanks identifying your organization's pandemic core team members by title, or specifically identify by title, spin-off crisis management team members, e.g., business closures crisis management team.

Title:

____

____

____

____

____

____

Similar to the core team, the ad hoc team members identified in the plan by function and not by name. These may include medical, distribution, insurance, information technology, among others. Please identify below possible ad hoc crisis team members for your organization.

Title:

____

____

____

____

____

____

Use a crisis management system in addressing a pandemic or crisis. Procedures that may be used include those from the World Health Organization, the US CDC, organization procedures, among others. All procedures should be referenced and their physical location identified in the plan. Actions to follow by the pandemic or crisis management teams at times of pandemic or crisis.

Plan

- The pandemic or crisis management team leader, who is the organization leader, calls the meeting to order. He or she identifies who is not present in the situation room, on the telephone, or participating via videoconference. Where people are not present, the facilitator is asked to provide information regarding the responsibilities of the missing people so they may be shared among those who are present, on the telephone, or participating via videoconference. Details concerning the situation room may be found in an earlier chapter.

Do

- At this point, the organization leader turns the meeting over to the facilitator to ensure that the pandemic or the crisis management plan is followed.
- If it has not already taken place, the facilitator asks the communication leader to have the communications sub-team begin drafting a standby statement. Draft statement provided in appendix 3.
- The facilitator leads the pandemic or crisis management team in determining which, if any at this point in time, of the ad hoc members of the team should be at asked to join.
- The pandemic or crisis management team needs to identify a scribe, a spokesperson, and a crisis manager. The scribe should identify and note all actions, new information, and the time when they were learned by the pandemic or crisis management teams. When the pandemic or crisis is reviewed at a later date, it is important to know what happened and when it happened. The writing of actions will not only provide a timeline but also the logic behind

decision-making. This may become important, as part of a learning process, when critiquing the pandemic and crisis management teams actions. The organization spokesperson is usually the head of the communications department. The crisis manager is usually an organization leader.

- A description of the pandemic or crisis management situation needs to be created and placed in front of a pandemic or crisis management teams so they remain focused on the pandemic or crisis and not pulled in various directions because of new information.
- The pandemic or crisis management teams will be responding to a pandemic or spin-off crisis by providing resources, creating messages, responding to the media questions, and anything else that is necessary to bring the pandemic or crisis to stability.
- Pandemic and/or the crisis management teams need to begin reviewing information received about the pandemic or spin-off crisis. They also need to begin reviewing outside resources that could be brought to provide advice via Internet, the telephone, or in person.
- The pandemic or crisis management team needs to review the standby statement if it has not already been issued.
- The speed of the pandemic or crisis should be reviewed by the respective teams to determine if it is slow, moderate, or rapidly developing crisis.
- Both the pandemic and crisis management teams need to remain focused so that they are not working on sub-crises that will delay reaching stability.
- Post-stability, the pandemic or spin-off crises teams may step back and put the responsibility of management and the hands of the pandemic manager and the crisis manager. The pandemic manager or the crisis manager may recall the pandemic management or the crises management teams if appropriate.
- The pandemic management or the crisis management teams may need to function around the clock or for a few

hours a day. Neither the pandemic management team nor the crisis management team should work more than twelve continuous hours. Alternates to both teams should take over if the pandemic or crisis goes beyond twelve hours and there's a need for the teams to continue working. If the pandemic crisis goes beyond twenty-four hours and there is a need to have the team continue working, the original team returns to take over. The retiring team should be nearby in a rapidly developing pandemic or crisis, perhaps sleeping on-site in a designated area.

Check

- *Record.* Records of activities associated with the pandemic and the spin-off crises are kept. Sometime in the future, there will be a critique of both the pandemic management team and the crisis management teams activities. It is a learning process where the teams will discuss what went right what went wrong and what was learned. It is obvious that it is important to conduct exercises of both the pandemic and crisis management teams to ensure a credible response. The record will be very helpful in the review. The record also includes who is trained in pandemic and crisis management and when their training occurred. In addition, the record includes who has been trained in working with the media.

- *Training.* Training requirements for both the pandemic and crisis management teams' core ad hoc members is important if they are speaking with the media or will be interviewed by the media. The core of the ad hoc members of each team must undergo pandemic and crisis training at least once every three years and partake in a crisis exercise at least once per year. If a team member will have an opportunity to speak with the media, they need to undergo media training.

Act

Documentation. Where new documents have been prepared and updated for both the pandemic and crisis management plans, they need to be approved by management, after which they become part of the organization's management documents.

- Appendices are used so that the pandemic and crisis management documents are not filled with unnecessary paper. For example: if there is a need to contact an expert in communications, the name of the contact phone number along with experts in other areas may be found in an appendix.

Please identify how your organization will do pandemic and crisis management document management, who will have access to the documents, and how updates will be done and approved.

APPENDIX 5

List of Organization Documents

Examples of organization documents for use during a pandemic or spin-off crisis management:

Number	Document Name	Document Location
1	Ethics Policy	Pandemic Facilitator's Office
2	Pandemic Management Plan	Pandemic Facilitator's Office
3	Spin-Off Crisis Mgt Plans	Pandemic Facilitator's Office
4	Pandemic and Crisis Training	Pandemic Facilitator's Office
5	Organization Organograms	Leader's Office
6	Organization Health Policy	Leader's Office
7	List of Government Contacts	Leader's Office
8	Others as appropriate	To be determined

About the Author

Dr. Marchesani received his PhD degree from Rutgers University in New Brunswick, New Jersey, and has a master's degree and bachelor's degree from Drexel University in Philadelphia, Pennsylvania.

While working in the chemical industry, Dr. Marchesani was instrumental in the design and implementation of environmental, health, and safety governance, management programs, leadership efforts, performance improvements, and crisis management systems. He is president and CEO of Environmental Health and Safety International LLC.

Dr. Marchesani has published many papers in both national and international journals, holds three copyrights, and contributed to the book *Air Pollution*, third edition, volume 8, by Dr. Arthur C Stern. Dr. Marchesani also published a book titled *The Fundamentals of Crisis Management*.

Dr. Marchesani and his wife have three children and live in Bonita Springs, Florida.